NORTH CAROLINA
STATE BOARD OF COMMUNITY COLLEGES
LIBRARIES
ASHEVILLE-BUNCOMBE TECHNICAL COMMUNITY COLLEGE

DISCARDED

JUN 30 2025

D1794405

Managing the Design-Manufacturing Process

McGraw-Hill Engineering and Technology Management Series
Michael K. Badawy, Ph.D., Editor in Chief

ENGLERT · *Winning at Technological Innovation*
ETTLIE AND STOLL · *Managing the Design-Manufacturing Process*
RITZ · *Total Engineering Project Management*
STEELE · *Managing Technology*
WHEELER · *Computers and Engineering Management*

For more information about other McGraw-Hill materials, call 1-800-2-MCGRAW in the United States. In other countries, call your nearest McGraw-Hill office.

Managing the Design-Manufacturing Process

John E. Ettlie, Ph.D.
University of Michigan
Ann Arbor, Michigan

Henry W. Stoll, Ph.D.
Square D Company
Palatine, Illinois

McGraw-Hill, Inc.
New York St. Louis San Francisco Auckland Bogotá
Caracas Hamburg Lisbon London Madrid
Mexico Milan Montreal New Delhi Paris
San Juan São Paulo Singapore
Sydney Tokyo Toronto

Library of Congress Cataloging-in-Publication Data

Ettlie, John E.
  Managing the design-manufacturing process / John E. Ettlie, Henry W. Stoll.
     p. cm.
  Includes bibliographical references.
  ISBN 0-07-019713-X
  1. Design, Industrial—Management.   2. Industrial engineering.
I. Stoll, Henry W. II. Title.
TS171.4.E87 1990   89-78532
658.5'752—dc20   CIP

Copyright © 1990 by McGraw-Hill, Inc. Printed in the United States of America. Except as permitted under the United States Copyright Act of 1976, no part of this publication may be reproduced or distributed in any form or by any means, or stored in a data base or retrieval system, without the prior written permission of the publisher.

1 2 3 4 5 6 7 8 9 0   DOC/DOC   9 5 4 3 2 1 0

ISBN   0-07-019713-X

The sponsoring editor for this book was Robert W. Hauserman, the editing supervisor was Valerie A. Rothlein, the designer was Naomi Auerbach, and the production supervisor was Dianne L. Walber. This book was set in Century Schoolbook. It was composed by McGraw-Hill's Professional and Reference Division Composition Unit.

Printed and bound by R. R. Donnelley & Sons Company.

> Information contained in this work has been obtained by McGraw-Hill, Inc., from sources believed to be reliable. However, neither McGraw-Hill nor its authors guarantees the accuracy or completeness of any information published herein and neither McGraw-Hill nor its authors shall be responsible for any errors, omissions, or damages arising out of use of this information. This work is published with the understanding that McGraw-Hill and its authors are supplying information but are not attempting to render engineering or other professional services. If such services are required, the assistance of an appropriate professional should be sought.

For more information about other McGraw-Hill materials, call 1-800-2-MCGRAW *in the United States. In other countries, call your nearest McGraw-Hill office.*

Contents

Contributors ix
Preface xi
Acknowledgments xiii

Part 1 Issues and Opportunities

Chapter 1. Managing the Design Process — 1

Introduction — 1
Disciplined Anticipation — 3
The B24 Liberator — 5
The 1955 Chevrolet: A "Fast Lane" Product Launch — 8
Managing the Innovation Process — 12
When It All Comes Together — 14
Summary — 20
References — 20

Chapter 2. Bridging the Culture of Engineers: Challenges in Organizing for Manufacturable Product Design — 21

Introduction — 21
Perspectives on Organizational Culture — 25
The Culture of Product Engineering — 31
The Culture of Manufacturing Engineers — 36
Perspectives on Managing the Design-Manufacturing Interface — 42
Summary — 49
Notes — 50
References — 51

Chapter 3. Methods That Work for Integrating Design and Manufacturing — 53

Introduction — 53

v

The Design-Manufacturing Paradox — 55
The Five Key Integrating Actions — 56
Outcomes of Successful Design-Manufacturing Integration — 57
Ratio of Manufacturing Engineers to Design Engineers — 58
Structuring For Design-Manufacturing Integration — 59
Satisfaction With the Process of Design-Manufacturing Integration in a Unit — 64
Successful Design-Manufacture Integration — 66
Summary — 68
Cautions for Reorganizing — 69
Appendix 3.1. — 70
References — 77

Chapter 4. Design for Life-Cycle Manufacturing — 79

Introduction — 79
Underlying Concepts — 80
Improving the Design Process — 89
DFM Approaches — 93
Summary — 112
References — 113

Part 2   Case Histories — 115

Chapter 5. Revitalizing the Manufacturing and Design of Mature Global Products — 117

Introduction — 117
Black & Decker — 118
The Results of Double Insulation — 125
Summary and Conclusions — 130
Appendix 5.1. Competitor Analysis by Sunbeam Appliance Co. — 130

Chapter 6. GM: The Quad-4 Engine — 133

Introduction — 133
The Product Program — 135
Suppliers — 139
The Role of Quality in the Buyoff — 144
Plant Organization — 144
Outcomes of the Quad-4 Program — 149
Appendix 6.1. BOC—Powertrain Delta Engine Supplier Survey — 153
References — 157

Chapter 7. GE: Product-Process Development Management — 159

Introduction — 159
Organizing for Change — 160
Reducing New Product Introduction Time — 170

The Project Approach	173
The Salisbury Project: The Story Behind the Story	181
Summary	184
References	185

Chapter 8. IBM Corporation: Early Manufacturing Involvement (EMI) — 187

Introduction	187
IBM Rochester, Minnesota	187
Early Manufacturing Involvement (EMI)	188
One-Pass Design	189
EMI Product Focus	191
EMI Process Focus	192
Manufacturing Designers	195
Conclusions	197

Chapter 9. A. B. Chance: Integration of the Design Process — 201

Introduction	201
Company Design History	202
Management of the Design Process	207
Problems, Results, and Expectations	212
Conclusion	220
References	221

Chapter 10. Northern Telecom: The Gate Procedure — 223

Introduction	223
Northern Telecom	224
Teamwork and Techniques: A Response to New Technological Challenges	226
Frequent, Short-Cycle Projects: The Gate Procedure Response	229
Gates and Gate Reviews	231
"Prime" Responsibility	233
The Stages of New Product Development	234
The Gate Procedure in Operation	236
The Gate Procedure: Management Tool or Replacement?	241
Conclusions	241

Chapter 11. Implementing Simultaneous Engineering at Cadillac — 243

Introduction	243
Create a Vision With Organizational Support	244
Develop Steering Committee	245
Analysis	246
Organization Design and Planning	246
Implementation	249

Development and Continuous Improvement 251
Learning 252

Part 3 Baseline for the Future 255

Chapter 12. Integrated Design Management 257

Introduction 257
New Principles 257
Economic Planning for Design 268
Summary 275
References 277

Index 279

Contributors

Klaus M. Blache *Advanced Engineering Staff, General Motors Technical Center, Warren, Mich.* (CHAP. 6)

Paul D. Coughlan *Lecturer, London Business School, London, England* (CHAP. 10)

John E. Ettlie *Director, Manufacturing Management Research, University of Michigan, Ann Arbor, Mich.* (CHAPS. 3 and 6)

H. Dennis Haubein *Vice President Engineering, A. B. Chance Company, Centralia, Mo.* (CHAP. 9)

James N. Hughes *Retired, formerly with Corporate Engineering and Manufacturing, General Electric, Bridgeport, Conn.* (CHAP. 7)

Robert F. Jones *Organization Research and Development Manager, General Motors Corporation, Detroit, Mich.* (CHAP. 11)

Alvin P. Lehnerd *Vice President, Product Development, Steelcase Inc., Corporate Development Center, Grand Rapids, Mich.* (CHAP. 5)

William H. Monsen *Development Engineering Manager, IBM Corporation, Rochester, Minn.* (CHAP. 8)

Steven Nowak *Graduate Research Assistant, School of Business Administration, University of Michigan, Ann Arbor, Mich.* (CHAP. 6)

Stephen R. Rosenthal *Professor of Operations Management, School of Management, Boston University, Boston, Mass.* (CHAP. 2)

Henry W. Stoll *Technical Director—Design Technology, Square D Company, Palatine, Ill.* (CHAP. 4)

Albert R. Wood *Professor, School of Business Administration, University of Western Ontario, London, Ontario* (CHAP. 10)

NOTE: Chapters not otherwise designated were jointly written by the authors, John E. Ettlie and Henry W. Stoll.

Preface

This is a book about managing design. The book is for people who have to live with design issues and who are motivated to improve the design process. It is not the final word. To us, it is the initial statement of what needs to be done. The value of this book is in the wide variety of experience that we summarize and the convergence of insights that are distilled. Our intent is to capture this experience by providing the reader with a *disciplined anticipation* of the force of strategic design.

Why did we write this book? We have seen a lot of bad designs—products that did not work, manufacturing systems that failed, and organizations that never fulfilled their destinies. We have seen the struggle, and more than that, we have participated in it. We know it is real and that it is important to those who make a living doing design. Most importantly, we have seen people endeavoring to implement bad advice that is masquerading as good advice.

The book is divided into three parts. Part 1 is an introduction to the problem and a specification of the need for a philosophical shift in design management. Part 2 describes seven case histories from industry experience that illustrate this philosophical shift. Part 3 summarizes what comes next.

Part 1 consists of the first four chapters. In Chap. 1 we show that manufacturing organizations are capable of good design and that many of the needed behaviors sought today were a way of life in the past. We show that there is an important relationship between organizational design and product-process system design. Chapter 2 takes up the issue of the two cultures most relevant to design issues: engineering and manufacturing. Next, we review survey results of a variety of practices for structuring the design process and the performance outcomes of these alternatives in Chap. 3. In Chap. 4, we introduce the philosophy of design for life-cycle manufacture.

Part 2 presents seven chapters that extract the case histories of companies that have come to grips with the philosophical shift in managing the design process. Chapter 5 is the chronology of Black & Decker and its response to competitive crisis through product-process design. The GM Quad-4 engine case history is described next in Chap. 6 where the role of supplier involvement and the shortening of the product introduction cycle are highlighted. Chapter 7 traces the pioneering efforts of GE and other firms to manage projects more effectively and structure organizations more efficiently. The IBM case history presented in Chap. 8 illustrates the innovative use of early manufacturing involvement (EMI) to design and launch intermediate range computers such as in the System/36 program. The A. B. Chance Company, which makes equipment for the electrical power industry, is included in Chap. 9 because of their novel approach to CAD/CAM integration and leveraging this approach to optimizing design. Chapter 10 presents the Northern Telecom case, which illustrates how the gating procedure can be modified for a specific product launch at a given plant site: the Harmony Telephone™ at the London, Ontario manufacturing facility. Finally, in Chap. 11, the unique perspective of an organizational development consultant is introduced by way of the case history of implementing simultaneous engineering at the Cadillac Car Division of GM.

Part 3 consists of the final chapter, which is both a summary and a forecast of what future-oriented manufacturing companies can do. This includes the use of collaboration and, in this chapter, we articulate the meaning of the philosophical shift in design that is illustrated throughout the book. We emphasize what is common or shared in all of the cases, the experience of designers and managers, and the results of surveys of the design process. We introduce costing and economic planning to design. In this chapter, we show that there is a bridge to the future for all manufacturing companies, and this bridge is strategic design.

John E. Ettlie
Henry W. Stoll

Acknowledgments

This book would not have been possible without our contributing authors. Not only were they willing to write or collaborate on chapters, they were also willing to work to a deadline. We are indebted to them for their persistence in working with us on this project. We hope that our practicing collaborators prosper through excellent designs, and that our academic colleagues reap accolades from the insights they provided. Our contributing authors were Stephen R. Rosenthal, Alvin P. Lehnerd, Steven Nowak, Klaus M. Blache, James N. Hughes, William H. Monsen, H. Dennis Haubein, Albert R. Wood, Paul D. Coughlan, and Robert F. Jones. We also had the special assistance of Herb Schneider at GE.

We need to thank the companies that allowed our industry collaborators to participate in this project. This contribution "in kind" gave us a rare view of the design process from a variety of perspectives. These companies are Black & Decker, IBM Corporation, GE, Northern Telecom, A. B. Chance, Cadillac Division of GM, and Square D Company.

Our original collaboration on this project began at the Industrial Technology Institute in Ann Arbor nearly four years ago. The supportive climate of that institution and of Donald Falkenburg, then President of ITI, are gratefully acknowledged.

A number of graduate students worked with us on design studies that prepared us for this book. Among them, Debbie Schut, Stacey Reifeis, Jeff Pozy, and Ernesto Reza were the most involved of our helpmates on this project. To all of our colleagues and mentors who make up a long list of very capable people, we say, "thanks for tolerating us." Several of these people read drafts of the book and made helpful comments: Paul Kleindorfer, Carol Tierney, Steve Rosenthal, and Dan Maas.

We thank our series editor, Michael Badaway, for having the confidence in us to promote the project and for making comments on the manuscript. We especially want to thank Bob Hauserman, the McGraw-Hill editor on the project, for his unwavering encouragement and problem-solving abilities.

We thank also Jackie Jeske of Square D Company for her able assistance with manuscript typing and word processing.

Finally, we want to thank our families for their tolerance of our absenteeism. They made the choice to write this book just as much as we did. Heaven help us without them.

Series Introduction

Technology is a key resource of profound importance for corporate profitability and growth. It also has enormous significance for the well-being of national economies as well as international competitiveness. Effective management of technology links engineering, science, and management disciplines to address the issues involved in the planning, development, and implementation of technological capabilities to shape and accomplish the strategic and operational objectives of an organization.

Management of technology involves the handling of technical activities in a broad spectrum of functional areas including basic research; applied research; development; design; construction, manufacturing, or operations; testing; maintenance; and technology transfer. In this sense, the concept of technology management is quite broad, since it covers not only R&D but also the management of product and process technologies. Viewed from that perspective, the management of technology is actually the practice of integrating technology strategy with business strategy in the company. This integration requires the deliberate coordination of the research, production, and service functions with the marketing, finance, and human resource functions of the firm.

That task calls for new managerial skills, techniques, styles, and ways of thinking. Providing executives, managers, and technical professionals with a systematic source of information to enable them to develop their knowledge and skills in managing technology is the challenge undertaken by this book series. The series will embody concise and practical treatments of specific topics within the broad area of engineering and technology management. The primary aim of the series is to provide a set of principles, concepts, tools, and techniques for those who wish to enhance their managerial skills and realize their potentials.

The series will provide readers with the information they must have and the skills they must acquire in order to sharpen their managerial performance and advance their careers. Authors contributing to the series are carefully selected for their expertise and experience. Although the series books will vary in subject matter as well as approach, one major feature will be common to all of them: a blend of practical applications and hands-on techniques supported by sound research and relevant theory.

The target audience for the series is quite broad. It includes engineers, scientists, and other technical professionals making the transition to management; entrepeneurs; technical managers and supervisors; upper-level executives; directors of engineering; people in R&D and other technology-related activities; corporate technical development managers and executives; continuing management education specialists; and students in engineering and technology management programs and related fields.

We hope that this series will become a primary source of information on the management of technology for practitioners, researchers, consultants, and students, and that it will help them become better managers and pursue the most rewarding professional careers.

MICHAEL K. BADAWY
Professor of Management of Technology
The R. B. Pamplin College of Business
Virginia Polytechnic Institute and State University
Falls Church, Virginia

Part 1

Issues and Opportunities

Chapter

1

Managing the Design Process

Introduction

The Aeronautical Systems Group at Lockheed Corporation recently developed an integrated metal-bending facility called Calfab. This mini-factory uses computer-aided layout and fabrication and has shortened the time it takes for design and manufacture of sheet metal parts from 52 days to 2 days—a 96 percent improvement. Metal used to travel 2500 ft between various machines and now it travels only 150 ft.*

There are few manufacturing firms left that have not targeted at least a 50 percent reduction in the time it takes to launch a new product from idea to production. Companies like Xerox have already accomplished this goal. Few organizations have pushed this concept to the point of having a corporate design strategy or a way of projecting the design and full-range planning of all their products five years into the future.† But this is coming. It is the rare company that has an innovation strategy that includes decisions about the business and new products, risk, and production.‡

Business Week had a very comprehensive article on this subject which included the material on Lockheed by Otis Port, et al., May 8, 1989, 142–150. There was also an article in the *New York Times* Business Section by John Holusha on July 8, 1989 where the case histories of Eastman Kodak, Timken and Corning were presented in detail with similar, impressive results.

†Corporate Design Strategy is introduced nicely in an article by Peter Lawrence in the Forum section of the *New York Times* Business section on February 12, 1989.

‡The *Journal of Product Innovation Management* had a special issue on Significant Issues for the Future, Vol. 1, 1984, 56–66. In this article, Merle Crawford introduces the concept of innovation strategy and talks about the four myths of new product introduction including the myth that all it takes is a good idea. Steven Wheelwright and Earl Sasser also have a nice article germane to this topic called "The New Product Develop-

Good ideas that are novel have a unique motivating quality. People get excited about them and eventually there will be competition and disagreement about their origin. Nonetheless, most ideas—good or bad—are never acted upon either by individuals, groups, and, especially, enterprises.

This book is about how new ideas get put into action. Our organizing theme is quite simple: it is both possible and desirable to manage the design process in manufacturing better. This is the process by which ideas are converted into results. One of the great myths of new product introduction is that all it takes is a good idea. As one of our colleagues, Merle Crawford, has said, "There is no such thing as a generically good idea," one that has found both its organizational mechanism for implementation and its market for success.*

Disciplined Anticipation

Much of what is in this book could be called *disciplined anticipation* of the consequences of having a good idea. Part of this concept involves knowing how ideas are converted from one form to another. For example, how the idea for a new product is realized in production and then becomes a complete delivery system is still a substantial challenge in most industries. To accomplish this disciplined anticipation, one will have to adopt a new approach toward thinking of a work organization as having "degrees of freedom" of operation and wondering how these degrees of freedom can be expanded during the unfolding of a project or program.

Although the challenge of adopting an integrated approach to the design process applies keenly to discrete parts manufacturing like automotive, aerospace, or appliance manufacturing, the realization that product, process, and service are inextricably connected raises the whole notion of how different functions must be optimally coordinated or integrated to accomplish a task. In a process industry like paper or steel treating, this might involve the process engineer and the software engineer or systems analyst. The operations, marketing, and information functions in banks are learning new ways of working together in firms like Wells Fargo.† There are lessons to be learned from all these cases.

We believe the unique adoption of current and leading-edge prac-

---

ment Map," in the *Harvard Business Review,* May–June, 1989, 112–125. Among other things they say "Product life cycles are short. Don't let families evolve one product at a time."

*This quote is also from Merle Crawford's section of the special issue of the *JPIM* (1984), on page 59.

†The Wells Fargo case appeared in the Business section of the *New York Times,* June 4, 1989, "Getting the Electronics Just Right."

tices for product-system development will only work for a firm if they are tailored to the local and regional cultures of that organization. Competitors can imitate the results of a design-for-assembly product, but they cannot easily imitate the culture that created the product or the next generation of new products and services of a firm. What is more, an integral part of the culture will be the history of the firm and its technologies defined in very broad terms as both hardware, software, and a knowledge base to effectively use these innovations. This history will promote or restrict the degrees of freedom available for planning better designs.

Are there general historical precedents for the new wave of simultaneous engineering and product management policies? We explore the answer to this question by presenting two historical case studies of unique product launches: the B24 Liberator, from World War II Ford Motor Company, and the 1955 Chevrolet, from post-war General Motors (GM).

The B24 Liberator

This project really got started in January 1939, well before World War II, when Consolidated Aircraft Corporation was asked by the government to design a new bomber. The bomber was sold on its wing design and took its maiden flight on December 29, 1939. What we're interested in here is the B24-C or third version design of the bomber that was "mass" produced at the Ford Willow Run plant in Ypsilanti, Michigan. This version of the B24 was redesigned for assembly and we thought it might reflect "modern" principles of Design for Manufacture (DFM). That is why it was selected.

In the process of developing the case, we relied on several sources including an article written by the then president of Consolidated Aircraft, R. H. Fleet (Fleet, 1943); and other articles (*Business Week*, 1943; *American Machinist,* 1943 and 1942; and Cackley, 1989). Finally, one of us interviewed Mr. Fred Klemach, who eventually became the chief aero engineer at the Willow Run Plant during the period the B24-C was in production.

The plan to mass produce the B24-C had some essential features that turned out to be difficult to implement, even though the concept eventually worked. This suggests that designs which anticipate assembly—even if they are fixed-up designs like the B24-C—might not necessarily reduce the lead time to launch production the first time they are tried. As a matter of fact, the Willow Run startup had so many problems in 1942 that the plant acquired the nickname "Will it run."

Willow Run was initially planned as a parts plant and did supply parts as well as planes. However, the essential feature of mass producing an aircraft—the prepunching of rivet holes, for ease and standardization of assembly—simply did not work at first. Many of the holes (about 20 percent) on sections did not line up. All assembly accessories were hung on sections before they arrived for final assembly so that people did not have to work inside the aircraft. The plant was criticized for not having the flexibility of a typical aircraft assembly operation, but it did make dozens of engineering changes daily (*Business Week*, 1943) and was preferred in the field by maintenance personnel because of the B24-C's interchangeability of parts (Cackley, 1989).

Willow Run was only able to produce 700 B24-Cs the first year of production (September 10, 1942 to September 1, 1943) but ramped up quickly to a production peak of 432 bombers per month in August 1944 (Cackley, 1989). The plant ran so well, it had to drop its third shift and run two 10-h shifts. The Willow Run Management Organization Chart is presented in Table 1.1. Note that Charles Lindberg was on the Ford payroll although his real role at Ford was not known until the end of the war. Ostensibly, Lindberg was a test pilot at Willow Run Airport. This turned out to be only partly true.

Engineers like Fred Klemach worked on the floor with assembly people and production supervision—making changes, solving problems when they came up, and meeting production targets within specifications. They even initiated a project to try and innovate around the "Davis" wing design and were able to show that part of the wing could

TABLE 1.1 Ford Motor Company: Willow Run Management (World War II)

Name	Title
Henry Ford	
Edsel Ford	
Charles Lindberg	
M. L. Bricker	Executive Vice President (Outlying Plants)
Charles Sorenson	Executive Vice President (Rouge Plant)
Roscoe Smith	Plant Manager at Willow Run
Logan Miller	Production Manager
William Pioch	Director of Engineering
Carl Scott	Chief Engineer
Wilson DeGroat	Chief Aero Engineer
Charles Koch	Ass't. Chief Aero Engineer
Serge Tchemesoff	Chief Aero Engineer
DeGroat replaced by Karl Scott	
Tchemesoff replaced by Fred Klemach	
Karl Scott replaced by Fred Klemach	

Source: Fred Klemach

be defined by rearranging the wing equation. It was the same as an older design. The rest could not be proved to be prior art, however, so Ford had to continue to pay royalties on each bomber produced.

Near the end of the war, Fred was asked by his boss what he was going to do when it was all over. Fred said he wanted to stay on at Ford if possible. So his boss took him to the archives and what appeared to be a library. In the back there was a bookcase with a lever door opener and it swung open. Behind the door/bookcase was a single room with a large table and numerous drawings laid out. It was Ford's project for the post-war aircraft industry—a huge, oversized transport. Fred began spending three or four hours a day working on the project in secret. If he was needed on the floor, someone who knew someone who knew someone could find him. Only a few people including Fred, his boss, Henry Ford senior, and Charles Lindberg knew about the project.

Several weeks later, flying back from a business trip, Fred's boss told him that when they returned to Willow Run they might not have jobs. "Why?" asked Fred. It turned out that Ford senior and Lindberg had a falling out as a result of an argument and the entire project was finished. Fred ended up being part of a team that started Ford research and development (R&D).

The lessons learned at Willow Run were simple.

- The concept of predrilling holes for rivets to mass produce B24-Cs was a good one—but you can only do one thing at a time and you can't rush it. A system produced the plane.
- The techniques of the auto industry could be applied to the aircraft assembly.
- Sometimes the essential feature of a product (like the wing of the B24) can be copied. But a new airplane can be introduced that will improve upon current design (Ford's planned post-war transport that never flew).
- It is not clear that anticipation of assembly alone is sufficient to shorten the cycle time to production.
- The R&D on projects doesn't need to be lost, even if a given product is never produced.

Another important lesson to be learned that is an indirect result of this case history is that once someone shows that something can be done, it puts pressure on the whole industry to follow. For example, the P-51 Mustang, the World War II fighter, was launched from "beginning of preliminary design to delivery of the first plane" in 120 days.* Product lead-time performance like this takes more than just a market, it also takes competition, and it takes industry specific experience.

The 1955 Chevrolet: A "Fast Lane" Product Launch

The 1955 Chevrolet was a car with a new body, new V-8 engine (the first V-8 in a modern Chevy), and a new chassis and frame, that is, a new platform. The car was launched from concept to pilot production in 24 months (summer of 1952 to summer of 1954) and dealers were fully stocked for customers by September of 1954 for the 1955 model year. The car, which has become a classic is pictured in Fig. 1.1.

How was it possible for GM and Chevrolet to do it? The engine alone should have taken a minimum of 5 years by today's standard. The GM

Figure 1.1 1955 Chevrolet Bel Air convertible. (*Courtesy of Chevrolet Division, GM.*)

---

*There is a good article on the P-51 Mustang from which this quote originates in *Aviation*, June, 1943, Vol. 42, 234–237. There is also a very informative but short article on engineering management for weapons development by George F. Metcalf in the *IRE Transactions on Engineering Management*, June 1957, 82–84. In this article, Mr. Metcalf says, "a major limitation in weapon development will be our ability to plant the programs, to organize the teams, to integrate their efforts and to measure their progress—that is to *manage creatively*." (page 82).

Quad-4 engine, which is presented in a later case (Chap. 6), was introduced concept-to-car in the record time of 48 months.

One of the authors interviewed Mr. Harry F. Barr for the book, and his account of the 1955 Chevrolet case provides real insight into some of what we may have forgotten about coordination between functions for a major product launch (Fig. 1.1). Mr. Barr was the assistant chief engineer for Chevrolet as part of the launch team for the 1955 Chevrolet. His expertise was in engines, and a monograph written by one of his staff, (Sanders, 1955) describes the unique features of the first V-8 engine for the modern Chevrolet. The 265 in^3 engine delivered 162 horsepower at 4400 r/min and incorporated several unique design features and production techniques. For example, the block used only 9 major and 3 minor cores in casting, while others of the day used as many as 22 cores to cast a V-8 block. The engine used a pressed forged steel crankshaft rather than an alloy dorn crankshaft because of the elasticity and specific gravity of the former, and because of the forging capacity of the firm (see Fig. 1.2).

At the time of the introduction of the 1955 Chevy, the division had 11 assembly plants, with one production engineer each reporting to a chief of production engineers. This engineering manager, in turn, reported to the chief engineer, who was Ed Cole at the time. Jim Premo was responsible for the body of the new car and the team got started in June of 1952.

Harry Barr had been instrumental in designing the revolutionary new engine at Cadillac for the 1949 model year. Much of the V-8 development activity in the industry at that time was driven by the assumption that high octane fuels used for World War II aircraft would become widely available after the war. This was not immediately the case, however, and some engine design plans that incorporated 10:1 compression ratios had to be scaled back to 8:1 (like the 265 V-8 at Chevrolet). Mr. Barr brought this V-8 engine technology with him to Chevrolet. Ed Cole had also come from Cadillac (tanks) to Chevrolet just one month earlier (May 1952) as chief engineer. When the project started, the manufacturing people led by Ed Kelly erected a sign in the engineering offices: "Anything you can design, Mr. Cole, we can make." The team was off to a fast start.

There was actually a V-8, 231 in^3, engine under design when the team was assembled in summer of 1952 and this engine design was scrapped in favor of the more innovative 265. What is, perhaps, even more fortunate was that a new engine plant was under construction in Flint at the time in anticipation of new V-8 engine releases. This provided the opportunity to experiment with new production technology at the same time as the new engine was being designed. One such innovation was tubular frame side rails. Stamped rocker arms was an-

10   Issues and Opportunities

Figure 1.2 1955 Chevrolet V-8 engine. (*Courtesy of Society of Automotive Engineers.*)

other. At that point, the only thing the design needed was hollow push rods to meet the lubrication specification. Rochester products had just accomplished that task, so all the pieces of the engine—always the critical path of a complete new car launch—were falling into place.

Tooling engineers often gathered at the drawing board in engine design to talk about what was feasible and desirable—and what was not. Unlike today, all design work was done in-house and the design board was the conference room for engineering coordination. Other innovations were incorporated in the drive train like Hotchkis drive which had been torque tube previously at Chevrolet.

In those days, the car divisions did not exchange information and acted very autonomously. When Oldsmobile people were asked by GM to visit Cadillac engine people in 1947, there were more than just a few eyebrows raised. Red flags went up all over in both divisions.

Technology was transferred when people were transferred. The only thing that carried over from earlier Chevrolets to the 1955 model was the transmission option—the 2-speed automatic power glide and the standard 3-speed manual transmission. The new things came from in-house creativity.

In the summer of 1954 about 100 pilot vehicles were produced by the supervision of the assembly plant and suggestions from this group were incorporated in the production launch. The windows of the dealers were papered over so no one could see the new model—it was a tightly held secret—until the car was unveiled in September 1954. Soon after, demand outstripped supply. Many people immediately called it the "little Cadillac" because of its looks—the grill and tail lights. Most were unaware of the Cadillac–Chevrolet technology connection under the hood.

Most of the major changes and innovations came about exclusively as a result of the approval actions of the engineering policy group at GM. The rest of the changes and actions after production release of drawings were done by production engineers reporting back to the manager of production engineering after coordination with manufacturing in the plants. There was only really one point of coordination. There were few people involved—there were only two levels of approval, unless there was disagreement. Fenders for the 1956 and 1957 Chevrolets were changed easily and gave the car a whole new look. The platform changed again in 1958. But *all* engineers reported to engineering. Production engineers in the plants were responsible for selling engineering ideas (drawings) to manufacturing. Only if there were real concerns would these come back to engineering at Chevrolet.

The real strength of this system was how quickly changes could be made, regardless of whether the change originated in engineering or manufacturing. All drawings went through one product engineer per plant. There was little hesitation in decision-making. If a change had to be made, it was made and the drawings were temporarily changed, then redone at a central coordinating point. Chevrolet staff redid the permanent drawings, sometimes after the changes had already been implemented in the plant. Cars were simpler and documentation was easier.

In summary, there appear to be several factors that facilitated a relatively quick launch of the 1955 model year Chevrolet in a total of 24 months from concept-to-pilot car:

- Transfer of technology was accomplished using transfer of engineering managers from Cadillac to Chevrolet for the V-8 engine—usually the time limiting factor in a complete new car launch.

- All design for the car was done in-house.
- The drawing board acted as the engineering conference room, exploiting group creativity.
- All engineering, including production engineering, reported to the chief engineer; and there was only one production (manufacturing) engineer per plant.
- Tooling engineers worked with design engineers early in the product life cycle.
- Fewer people and fewer organizational levels were needed to make critical, important, innovative decisions or changes if they were needed.
- The team working on the project was product oriented—the goal was to launch a completely new car, nothing more, nothing less. There was no additional purpose, for example, like learning about some new technology.
- Manufacturing engineers (called production engineers then) were charged with selling manufacturing plant people on engineering drawings.
- Engineering design changes could result from either manufacturing or engineering equally well and quickly.
- Both a new engine plant (Flint) and a new product were under development at the same time.

Managing the Innovation Process

If one takes the broader view of the product development process and applies it to the manufacturing design challenge, some interesting insights emerge. It is not obvious what managing the innovation process means to a product innovator and a process innovator in a modern firm today. Let us trace from the general to the specific to see how these insights might be derived from this perspective.

We start by presenting the results of a project by Axel Johne and Patricia Snelson of the United Kingdom.*. They studied product development procedures in the United Kingdom and the United States and found there were definite differences between the successful product innovators and unsuccessful firms, regardless of country of origin. The results are summarized in two combined tables from their article

---

*The citation for this article is Axel Johne and Patricia Snelson, "Auditing product innovation activities in manufacturing firms," R&D Management, Vol. 18, 1988, 227–233. The authors state that "supportive top management and efficient interfunctional teamwork emerge as key factors...." and suggest that their Table 3 be used as the audit tool.

(see Table 1.2). They divide their innovators into two categories—"leading product innovators" and "less successful product innovators." They compare these two types of firms on seven characteristics including strategy, values, structure, and systems. Their results are heavily focused on top or general management behavior and attitudes. *Explicit plans* and budgets for rejuvenating old products, and *broad objectives* for new products are practiced by successful firms.

Less successful firms do not integrate strategies for old product planning and catch-up is typical for new products. In these firms, project managers of new ventures need the endorsement of the top and teams cannot function unless new structure lets them operate outside traditional lines of authority. The over-the-wall approach to old product changes and no approach at all under the systems category for less successful firms is still all too typical in both countries (see Table 1.2).

The application of these general observations is illustrated in the results reported in another very recent study on the barriers to implementation of Computer-Aided Design/Computer-Aided Manufacturing (CAD/CAM) by Carol Beatty and John Gordon.* We reproduce their table which summarizes the results of their work by examination of three types of barriers: structural, human, and technical (Table 1.3).

The structural barriers to implementing CAD/CAM include the wrong justification criteria (labor costs), lack of measures of benefits of all types—tangible and intangible, and lack of coordination. These are all topics we will take up later. The most important human barrier is not resistance to change but *loss of control* of the work situation. This replicates the best work published to date on the topic.†

One of the issues raised here is the lack of valid predictive measures of the down-stream outcomes. In a later chapter we review what domestic plant managers and staff said they are using to evaluate design-manufacturing integration with similar results: the *majority* say they only *use ultimate bottom line* results like cost and quality. When they do use in-process evaluation, part and fixture reduction is the typical audit metric. In Daniel Whitney's recent article that appeared in the *Harvard Business Review*‡ he observes that this is what good firms do: they analyze products in order to reduce the number of

---

*The Carol A. Beatty and John R. M. Gordon article entitled, "Barriers to the Implementation of CAD/CAM Systems" appeared in *Sloan Management Review*, Summer 1988, Vol. 29, No. 4, 25–33. These two authors based their model and conclusions on actual visits to both CAD and CAM implementing firms, primarily in Canada.

†The August 1989 special issue of *IEEE Transactions on Engineering Management* (Vol. 36, No. 3) on "The Social and Organizational Dimensions of Computer Aided Design," represents a good overview of these issues.

‡Daniel E. Whitney's article entitled, "Manufacturing by Design," appeared in the *Harvard Business Review*, July–August, 1988, 83–91. It includes material relevant to the issue of corporate design strategy formulation.

TABLE 1.2a Summary Characteristics of Leading Product Innovators

	For planning and executing:	
	Old product development	New product development
1. Strategy	Top management determines explicit plans and budgets for development work	Top management sets broad objectives for organic growth.
2. Shared values	Top management fosters understanding of the need for product evolution.	Top management fosters understanding of the need for really new products.
3. Style	Top management is supportive but does not meddle in development projects. Progress is checked regularly.	Top management is intimately involved, often on a day-to-day basis.
4. Structure	Top management uses the existing organization which acknowledges the need to manage updates within a matrix of responsibilities.	Top management uses new organizational forms such as business teams to nurture important developments outside the mainstream organization.
5. Skills	There is efficient product planning using sophisticated market analysis techniques.	Techno-commercial ideas generation, screening, and testing in concept. Development work often based on new technology.
6. Staff	Existing line managers are used with some staff advice. When project leaders are appointed they may be quite junior but receive a commission from top management.	An intrapreneur is allowed to select his/her own team with whom rewards are shared. Failures are viewed as a learning experience.
7. Systems	Loose-tight using simultaneous or rugby approach. More tight than loose.	Loose-tight using simultaneous or rugby approach. More loose than tight.

SOURCE: Axel Johne and Patricia Snelson. Reprinted by permission.

parts and migrate to jigless fixturing in manufacturing. His example of a radiator made without jigs is reproduced here in Fig. 1.3.

When It All Comes Together

Although the outcomes of this transformation process to simplified product design-product production systems might seem obvious, how one obtains a unique, culturally relevant solution to make this happen is not obvious. Boeing has been in the news a lot lately, but most of

TABLE 1.2b **Summary Characteristics of Less Successful Innovators**

	For planning and executing:	
	Old product development	New product development
1. Strategy	The product development strategy is not integrated within the corporate strategy: top management delegates responsibility for product development to the technical or marketing functions.	Top management prefers to catch up not organically but by acquistion
2. Shared values	Top management gives the responsibility for initiatives to mainly one function which then has to battle against the "not-invented-here" symdrome.	Top management accepts that the existing organization will feel threatened by any radical product developments which might undermine existing power bases.
3. Style	Top management either distances itself from or meddles in development projects.	Top management involves itself directly only in acquisitions. Any organic new ventures are left to individuals who will take the considerable personal risks involved.
4. Structure	Top management uses the existing formal structure supplemented with low-level staff specialists.	Top management undertakes any radical developments right outside the existing formal structure that is not organized new product development.
5. Skills	There is little or no explicit product planning. Sophisticated market analysis techniques are used rarely.	There is only a very limited in-house base for the skills needed because these skills have never been nurtured.
6. Staff	Staff specialists battle it out with existing line managers. When teams are formed they are led by a senior person.	Great reliance on outside consultants, or in the case of an internal "godfather".
7. Systems	Loose or tight: 'over the wall' approach to team-work.	None

Source: Axel Johne and Patricia Snelson. Reprinted by permisison.

TABLE 1.3 Barriers, Causes, and Remedies in Implementing New Technologies

Barrier	Causes	Remedies
Structural		
Excessive focus on direct labor and ratios	Obsolete decision criteria	Careful analysis of real costs and benefits
Failure to perceive true benefits	Lack of measures of intangible benefits	Analysis of total productivity and "intangibles"
High risk for managers	Reward systems discourage risk taking	Different reward systems for managers
Lack of coordination and cooperation	Organizational fragmentation	Devices to integrate and coordinate
High hopes and hidden costs	Overselling	Planning strategic objectives
Human		
Uncertainty avoidance	Fear of change and uncertainty	Involvement and communication
Resistance	Fear of loss of power and status	Careful implementation, champion
Hasty decisions and chronic fire fighting	Action orientation: impatience with planning and waiting	Preimplementation planning, long-term objectives
Technical		
Incompatibility of systems	Purchase of a variety of hardware and software	Buy only an integrated system, write your own software, neutral files

SOURCE: Reprinted from "Barriers to the Implementation of CAD/CAM Systems" by Carol Beatty and John M. Gordon, *Sloan Management Review,* vol. 29, no. 4, Summer 1988, pp. 25–33.

what Boeing does that is germane to this book may never be seen by public media sources. That is too bad. Here are some points that are worth noting.

In order to ensure that trouble-shooting gets done on new aircraft products, Boeing uses "liaison engineers" that deal directly with workers on the line to solve problems together when they come up. This is especially true for the Boeing 757, which first rolled out in 1982.* We have found in our own study of over three dozen moderniz-

---

*The material on "liaison engineers" appeared in an article by Roy J. Harris, Jr., titled, "As Boeing Focuses on Fire-Control Flaws, Safety Issues Widen," in the *Wall Street Journal,* April 11, 1989, on pages A1 and A10.

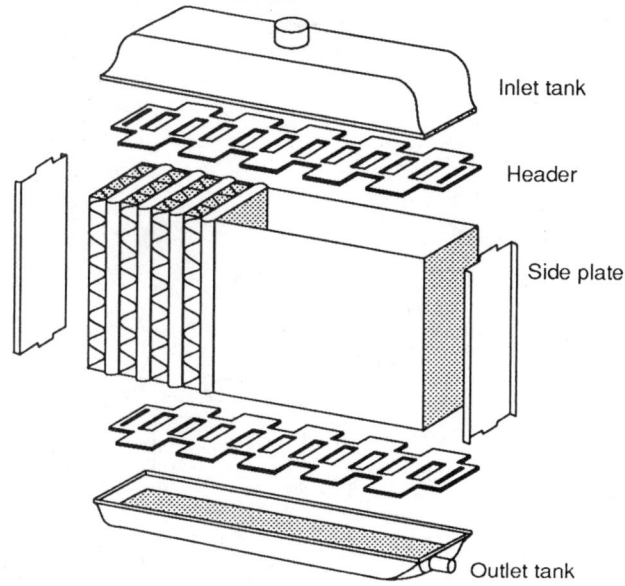

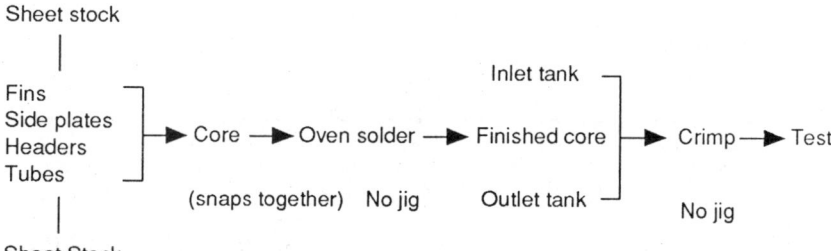

Figure 1.3 How a radiator is made—the combinatorial, jigless method. (Reprinted, by permission, from Harvard Business Review. Exhibit from "Manufacturing by Design," D. E. Whitnzy, July/August 1988.)

ing, domestic plants that one of the single best predictors of innovativeness is whether or not these plants have real (not imaginary) engineer blue-collar teams.

Again, from Boeing comes the adaptation of "design/build teams," outlined in a paper by Jim Krampert, an engineer with Boeing.† A summary chart (see Fig. 1.4) from his paper shows that these teams are really a hybrid combination of matrix and project organization

---

†Jim Krampert's talk "Using Design-Build Teams in Software Development," was given at the TIMS/ORSA meeting in the Vancouver, Canada, meeting, May 9, 1989, and he kindly supplied a reprint of the talk to us.

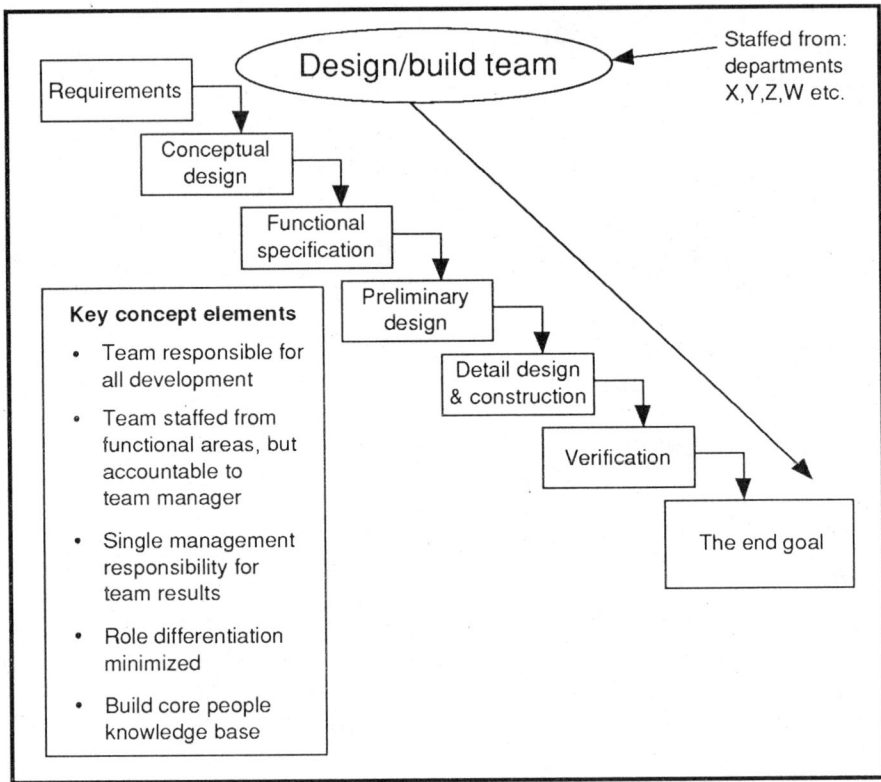

Figure 1.4 The design/build team concept. (*Reprinted, by permission, from James Krampert, TIMS/ORSA meeting, Vancouver, Canada, May 9, 1989.*)

used to integrate design, manufacturing, information, and other functions across project stages. Key concepts include *team delegated authority* in areas such as functional specification and management of role-differentiation (Fig. 1.4).

This brings us to action planning and techniques that might be applied to result in the conversion process of old to new management of the design process. We focus on the issue of conflict resolution. One of the most provocative methods being used to develop new product development policies is described by Ron Ebert and his colleagues in a recent case study of the Handy Company.* Ebert, et al. use a method called "judgment capture" to expose various operating assumptions

---

*The excellent article by Ronald J. Ebert, Clyde D. Majerus and Dale E. Rude, is titled, "Product Development: Assessing the Consistency of Engineering Design Policies," and it appeared as a technical note in *IEEE Transactions on Engineering Management*, (Vol. 36, No. 2) May 1989, 140–146. The judgment capture software is available separately.

TABLE 1.4 Factors and Policy Levels for Product Development

Development factor	Policy level		
	Low (1)	Medium (2)	High (3)
1. Goal/Deadline Emphasis	Weak	Moderate	Strong
2. Inter-group Information Transfer	No deliberate information transfer	Status update meetings	Full time project coordinator
3. Product Development Method	Individual assignments	Team assignments	Group development
4. Product Unit Cost Emphasis in Design	Unimportant	Medium emphasis	Very important (lowest cost)
5. Product Quality Emphasis in Design	Satisfactory product	Above average product	Excellent product

SOURCE: Reprinted from Ronald J. Ebert, et al., "Product Development: Assessing the Consistency of Engineering Design Policies," *IEEE Transactions on Engineering Management*, vol. 30, no. 2, May 1989, pp. 140-146.

and practices among various actors and groups that influence the design process that operate at different policy levels. The five development factors are then evaluated for these three policy levels and their results are summarized in Table 1.4.

The rationale for this approach is quite simple. In order to resolve differences among various value positions, these communication assumptions that can be influenced by policy need to be brought out into the open and confronted. This entire area of research has been labeled controversy theory and is reviewed elsewhere.† This does not assume that anything will change automatically. But the judgment capture software can at least operationalize various judgments into judgment scenarios. These can be evaluated by participants in the exercise. This documented scenario approach can then be used to facilitate a more coordinated product development process.

The attractiveness of the approach also lies in the fact that since the innovation process is inherently disruptive, it naturally creates conflicts among various parts of any "normal" organization. As Michael Tushman and David Nalder have observed, each of these groups has its own perceptions and priorities. Some conflict must be valued by the

---

†Controversy and open debate of opposing views was reviewed recently in an article by Dean Tjosvold, "Implications of Controversy Research for Management," *Journal of Management*, Vol. 11, No. 3, 1985, 4–37.

innovative organization to be successful and some method must be installed to deal with this "natural" conflict. The greater the learning required, the more important the informal organization to this process.*

Summary

Revamping the product management process in any firm is a challenge. Remaking an organization into a successful product-process-service enterprise is even more daunting. But firms are experimenting with new methods to make this happen and results are now beginning to be made public. It can be done.

This book is about the various ways transformation can be accomplished. It is not the final solution. Every organization must make its own way in the world with its unique culture of strengths and weaknesses as the state-of-the-art changes. Competitors change the world we live in and customers change the lives we lead. However, when you start somewhere, you will get someplace. This book is our recommended starting place.

References

American Machinist, "Tooling for Bombers at Willow Run," December 10, 1942, pp. 1441–1456.
American Machinist, "Ford Applies Mass Production to Big Bombers," February 4, 1943, pp. 93–108.
Business Week, "Willow Run Runs," February 20, 1943, pp. 17–18.
Cackley, Phil, "Looking Back at the Liberator," *Ann Arbor News,* May 16, 1989, pp. D1–D2.
Fleet, Major R. H., "Building the B24," *Aviation,* January 1943, pp. 66–69, 222–228.
Johne, Axel and Patricia Snelson, "Auditing Product Innovation Activities in Manufacturing Firms," *R&D Management,* vol. 18, no. 3, 1988.
Sanders, R. F., "The New Chevrolet V-8 Engine" SAE Golden Anniversary Annual Meeting, Sheraton-Cadillac Hotel and Hotel Statler, Detroit, Mich., January 10–14, 1955.

---

*Michael Tushman and David Nadler published their article entitled "Organizing for Innovation," in the *California Management Review,* vol. XXVIII, no. 3, Spring 1986, 74–92. This is an excellent introduction to the perspective that the research and management tradition in the technology management area can provide to the design community. But it is only a start, and readers should go beyond this if they are really interested.

Chapter

2

Bridging the Cultures of Engineers: Challenges in Organizing for Manufacturable Product Design

Stephen R. Rosenthal*

Introduction

Increased global competition throughout the 1980s has strengthened the significance of a company's ability to introduce new products. In what has been recognized as increasingly dynamic markets, customers' needs vary and shift more rapidly than has been traditional. Companies in industries including office equipment, consumer products, and machine tools responded by attempting to shorten the time required to design, develop, and manufacture new products, and by placing an increased emphasis on cost reduction, quality, and satisfying customer needs. The adoption of advanced technologies, such as computer-aided design (CAD) or various forms of flexible manufacturing systems, was a partial response in that these technologies offered the promise of "economies of scope" in manufacturing. Nevertheless, such technologies did not ensure a timely and successful new product introduction. Even if individual steps of a new product introduction effort—such as revising the design of a product or a component, planning the production process, testing a prototype, or "ramping-up" to

---

*The author acknowledges the helpful comments and suggestions received from John E. Ettlie, Merrill L. Ebner, and Mohan V. Patikonda in their review of this chapter.

full production volumes—are shortened, the product introduction may still falter.

Product development is inherently a creative, interactive problem-solving process, not just a technological accomplishment. Many companies realized that their traditional, sequential approaches to new product design, development, and manufacture had become a barrier to their competitiveness. It simply took too long and cost too much by the time a new, high-quality manufactured product was available in large enough volumes. The need for more coordination and collaboration among people who design products and those who must manufacture them had become apparent.

As some U.S. manufacturers began to experiment with more interactive processes for developing new products, terms like "concurrent engineering," "simultaneous engineering," and "early manufacturing involvement" were coined to describe the new approaches. These terms were rapidly diffused to many executive suites, as the business press (*Business Week*, 1989, for example) publicized the impressive accomplishments of such industrial "blue chips" as IBM Corporation, Ford Motor Co., Deere, 3M, Black & Decker, Hewlett-Packard, Lockheed Corp., and Northrop Corp.

Upon closer examination, however, the problem of managing the design process in manufacturing was far from being solved. Different companies were taking different paths for different reasons with different results. Consider the following cases in which three companies recently attempted to introduce a new product under extreme competitive pressure:*

Company A: Recently formed to develop a product that would compete in a rapidly growing and highly competitive market for electronic equipment. Its founder selected the key engineering and manufacturing managers with a careful eye to their willingness and ability to work together. In their prior jobs, these managers had experienced the problems of trying to introduce new products when independent functional orientations were more dominant. They joined Company A not only for its significant commercial potential but also for its special professional climate. From the beginning, Company A's corporate culture emphasized teamwork and integration of work activities in the design of the new product, the selection of sup-

---

*These cases are drawn from a current research project on "Product Design and Manufacturing Management," sponsored by the Boston University Manufacturing Roundtable. The project (which began in the Spring of 1989 and is scheduled for completion in the Fall of 1990) includes in-depth investigations of new product introductions in a variety of industries, with special attention to the role of manufacturing management. For further information on this project and its publications, contact the author at The Manufacturing Roundtable, Boston University, 621 Commonwealth Avenue, Boston, MA 02215.

pliers, and the development of a manufacturing capability. The founder of Company A believes that "business has gotten so complex and dangerous and quick-moving that no matter how brilliant people are at the top, it's not good enough any more. You have to utilize every brain in the whole organization." Company A seems to thrive on openness. Everyone speaks his or her mind in group problem-solving meetings. Everyone is committed to the design, development, and delivery of a top quality product. To a large extent the product and the manufacturing process to produce that product were developed simultaneously. From the outset, manufacturing involvement in product design decisions was the rule, not the exception. The founder of Company A strongly believed in the importance of manufacturing as a competitive weapon. This company hopes never to have problems with its design-manufacturing interface, although they recognize that if their plans for rapid growth succeed, they will have to be diligent to avoid the onset of rigid and conflicting functional cultures.

Company B: A well-established business in the telecommunications industry when it faced a special challenge calling for radical change in their process of new product introduction. An early, narrow market window appeared when one of the major competitors introduced a product that lacked significant advances and the other most aggressive competitor announced a delay in its next product offering. Company B set out to develop a piece of end-user equipment in a business where they had little former success; their reputation for quality products was based largely on more complex integrated electronic systems. Either a winning new product had to be developed and introduced quickly or an existing plant that was extremely underutilized would have to be shut down with a great chance that the engineering division would be shut down also. Facing strong time-based competition, they were able to design, develop, and manufacture this new product with no problems of quality in what was their record time of little more than one year. This product soon became a market leader for its functionality, flexibility, and quality. When launched, it had already achieved its "steady-state" cost target and profit margins were higher than traditional for that business. Senior managers at Company B credited their success to early collaboration among specialists in marketing, engineering, manufacturing, and testing. Manufacturability and testability was built in at the design phase and the manufacturing participants influenced the selection of product technology. Very few subsequent design changes were required. To accomplish all of this, Company B had to achieve remarkable changes in the cultures of both its engineers and its manufacturing experts for this was a very different new product introduction process from that to which they were all accustomed. In fact, the first people assigned to this new product introduction were reluctant to join the project. Participants, subsequently, gave credit to the strong new manager who was hired to run this project for his initiative in establishing a positive attitude and momentum, thereby discouraging potential resistance to the sudden change. One observer in the company likened the product development team to a "daytime family." Company B is now attempting to sustain and incrementally improve the design-manufacturing interface in subsequent new product introductions.

Company C: Well established in its market for electro-optical equipment when it was suddenly leapfrogged by a competitor who introduced a better and less expensive product. Though it was still fairly young and small, Company C had enjoyed rapid growth over the prior few years and was proud to be a leader in its industry. Their success had been based on allowing product engineers to retain a dominant role in defining the scope and timing of technological innovation in new products. In particular, the timing of development and manufacturing ramp-up were thought to be secondary to getting the best technology into the product. Now Company C had to try to respond in a more time critical manner. They could not afford the extensive process of engineering change orders (ECOs) that had characterized past product introductions, because responding to all of these ECOs delayed manufacturing's ramp-up to full production volumes. The Company organized a special task force and made a conscious attempt to have product engineers and manufacturing experts work together from the start on a more equal footing. Product designers had to get the early involvement of manufacturing experts (and software developers) before designs were "frozen." A new emphasis was placed on designing parts to use existing castings from previous products, thereby saving tooling costs; prior to this design engineers thought that their job was always to design totally new parts. Following these procedures, the ECO syndrome was largely eliminated. The new product was designed, developed, and manufactured in a shorter period of time and yet it did not fit target windows. Still, it is a tremendously successful product. Although quality and market acceptance were reasonable, they were not good enough to overcome the competitive threat that already existed. One key manager summed up the experience as follows: "We have lowered the wall between engineering and manufacturing, but we have yet to remove it." Company C attempted to learn from this experience by establishing a temporary task force to review this new product introduction. Shortcomings were noted and lessons for the future were formulated. The company is now trying to improve its new product introduction process even further, with full recognition that there are still cultural barriers to be overcome.

These three vignettes illustrate typical recent experiences of companies attempting to build or change the culture within which they design and manufacture products. Facing its own start-up, the founders of Company A built the desired culture from the outset. This was accomplished through the deliberate recruitment of competent technical specialists and managers all of whom wanted to work in a dynamic, integrated problem-solving climate devoid of traditional functional barriers. Company B, already established in its industry, launched a special project with a more interactive design-manufacturing interface. By so doing, they were able to turn a pending plant closing into a growth situation. Company C tried to do the same, although in a more informal and incremental way. Senior managers at Company C feel that their success in this instance was only partial and they are still working hard to avoid rigid functional approaches to new product introductions.

Each of these manufacturing companies grappled with a standard set of challenges in their continuing quest to build a more salable product. The design-manufacturing interface is critical to this goal since it is here that so much of the new product's cost, quality, and time to market can be affected, for better or worse. Conventional wisdom has shifted sharply away from faith in quick fixes; much of the decade of the 1980s was characterized by high enthusiasm for rapidly instituted, top-down quality "programs," or short honeymoons with computer-integrated-manufacturing (CIM) technologies. New prescriptions call instead for fundamental organizational and behavioral change in areas of new product introduction.

Perspectives on Organizational Culture

"Culture" is now recognized by managers in companies of all sizes and in all industries as either a major barrier or a potential opportunity for achieving an enhanced design-manufacturing interface. Unfortunately, at this point, there tends to be more consensus on *what* needs to be accomplished than on *how* to do it and what the pitfalls might be. The risks are particularly high when changes require new behaviors and organizational structures, when decisions deal with complex technological choice, and when organizational inertia is a dominant factor. The nature of this challenge is the subject of this chapter.

Implementing such organizational change requires product engineers (often called "design engineers") to work more closely and effectively with process engineers (often called "manufacturing engineers"). New organizational initiatives are often called for. The extent of change varies and might be aimed at requiring manufacturing sign off before product designs are released, introducing a special liaison person who can serve as a working buffer between designers and manufacturing engineers, using cross-functional teams, or creating a single department responsible for both product and process design (Dean and Susman, 1989). In selecting from the various possible organizational changes, management must understand and consider existing differences in culture between their design engineers and their manufacturing engineers (see Note 1 at end of chapter). Bridging these cultures or, better yet, developing a new integrated one is essential for significantly improving the design-manufacturing interface.

An article by Jay Barney (1986) helps set the stage for our investigation. Noting the existing debate in the literature, Barney adopts the commonly used notion that organizational culture is "a complex set of values, beliefs, assumptions and symbols that define the way in which a firm conducts business." He then goes on to raise and answer the question: Can organizational culture be a source of sustained corporate advantage? Barney argues that this can only occur when the particular culture is both rare and imperfectly imitable, as well as being valuable. "Valuable," to Barney, means that the firm's culture enables its creation of additional financial value. Presumably aspects of such a culture facilitate the type of changes whereby employees become better at designing, manufacturing, and marketing new products. "Rare" means that few, if any, of its competitors have a similar culture because of unique experiences either at its founding or during its subsequent growth. Finally, "imperfectly imitable" means that competitors cannot simply adopt those components of this valuable culture that they lack. Barney notes that this condition can arise because many such components are unspoken and hard to describe or because they arose through some unique aspect of the company's history. In any event, the fact that a company's culture already facilitates an effective design process in manufacturing does not imply that a sustained competitive advantage exists. Alternatively, where a company's current culture creates a partial barrier to effective new product development, an attempt to modify it to replicate the "successful" culture of another firm may or may not be possible. Indeed some organizational cultures may be hard to change to any significant extent.

Stanley Davis (1984) has written about corporate culture from the perspective of how to manage it. He makes a fundamental distinction between "guiding beliefs" and "daily beliefs." Guiding beliefs provide principles at a rather philosophical level; they deal with the best way to compete and how to manage a company. Such guiding beliefs, which rarely change, give direction to daily beliefs, which are more situational. Davis refers to daily beliefs as "the survival kit for the individual." While guiding beliefs support the formulation of corporate strategy, daily beliefs support the implementation of strategy. Davis uses the term "daily beliefs" to encompass how individuals ought to behave with respect to innovation, decision-making, communicating, organizing, monitoring, appraising, and rewarding.

At the corporate level, such considerations of culture, even the daily variety, are quite abstract. Studies of organizational culture tend to concentrate on the overall tone for success in a company, how the company views its basic mission, and the function of common values and principles in guiding the actions of employees. While this literature is somewhat suggestive for our purposes, it is far from conclusive. It is not anchored in the specific context of any particular kind of professional work. It does not address the notion of corporate subcultures, which have their own guiding beliefs, strategies, and daily beliefs. In short, it offers little direct guidance on issues like how to improve the design-manufacturing interface.

Our use of the term "culture," in contrast, is at the level of functional working groups, be they design engineers, manufacturing engineers, or some combination of the two. The literature on the culture of engineering practice is indeed limited (see Note 2 at end of chapter). Although there is a notable tradition in the field of organizational behavior of research on the climate of work groups (Schneider, 1985), we are unaware of any such systematic comparative studies involving product design and manufacturing in the management literature (see Note 3 at end of chapter). Researchers who have looked at these particular functions in recent years have looked at the productivity impacts of standard work practices rather than at cultural dimensions of work in such groups.

Viewing engineers as professionals, this section discusses differences between their general orientations and those of their managers, including typical attributes of such workplace cultures, and the legacy of their educational background. Subsequent sections of this chapter deal more specifically with cultural traditions and challenges first in the field of design engineering and then in manufacturing engineering. Finally, we focus on cultural gaps and approaches to resolving them, drawing on some classic concepts in the management literature.

Professional and managerial orientations to work

Product design and manufacturing engineering organizations experience the natural clash of cultures between any group of managers and salaried professionals. The traditional "command and control" dichotomy is often a stimulus to this clash between managers and technical professionals within a company. Management creates the structure within which problem solving takes place, identifies the problems to be worked on, and reviews the recommended solutions. Engineers, in turn, perform the analysis and develop the recommendations. In such a structure managers naturally orient themselves to the process of command and control, while professionals are geared to the solution of technical problems. In the more successful companies these two orientations would not generate debilitating conflict. In others they do.

Within a company, decisions must therefore be made about the proper amount of autonomy that individual (and groups of) engineers ought to have. Large U.S. corporations have tended to develop "chimneys of power" around their various specialized technical tasks. According to this vertical imagery, specialists work on small, contained projects which are integrated through a hierarchy of managerial coordination and review. If technical solutions are acceptable to other specialists and managers, new assignments are made. If they are not, the problem is escalated to a higher level of command and control, where other specialists can eventually add their perspectives in a forum that tends to pit one organizational group against another. In the design of many modern, complex electronic products, hardware experts may propose that a particular mechanism be designed in a certain way. But this way may require that the associated software, which is to be developed by another design group, be unacceptably complex and burdensome. Or perhaps the proposed mechanical design contains so many parts that the manufacturing engineer is concerned about the cost to assemble the various components.

Now consider what happens at a company that employs a variety of professionals, such as engineers. Here Joseph Raelin (1985) points to the dual problem of overspecialization and overprofessionalization and its associated challenges to corporate management. Over-specialization is a common characteristic within groups of design engineers or manufacturing engineers, because each field is composed of a number of well-accepted subfields. For example, electrical engineers may be computer engineers, power engineers, or control engineers. Overprofessionalization, in turn, is characterized by an exclusive orientation to one's special skills or knowledge apart from the broader goals of the organization: the proverbial "wall" between the worlds of product design and manufacturing. Despite any attempts to broaden the range of capabilities of its engineering workforce, the managerial challenge inherent in dealing with long traditions of specialization and professionalization remains: finding effective ways to pursue current business goals in part through the efforts of professionals with limited perspectives who conduct specialized tasks using technical skills.

Professionalization in design and manufacturing has traditionally meant that individual specialists desire and expect freedom to do the work that they know how to do best. Regardless of their own particular specialties, engineers as individual professionals are likely to share the common cultural norm of respecting others for their own unique contributions. Although they are likely to have opinions as to the relative competence among members of another specialization, most individual engineers can be expected readily to acknowledge the importance of those who work in specialties other than their own. Of more personal concern than the number and variety of other specialties, then, is the work that any one specialist gets to do. As pointed out by Badawy (1971) engineers are normally highly motivated to work on important problems and are likely to be frustrated to the extent that their company does not adequately use the professional skills and experience that they have acquired. This tends to exacerbate the clash between problem-finding professionals who desire change and manufacturing managers who have to satisfy customers' current orders and maintain organizational stability.

Professionals and their workplace culture

Such aspects of professionalism must be viewed in the context of workplace culture if we are to begin to understand the challenges for management in promoting effective product design and manufacturing. Rigid, functionally-dominated organizational structures naturally lead one group of engineers to feel superior to another group based on the perceived importance of their collective contribution to the company's well-being. Everyone's organizational base becomes, in a sense, the center of the universe. Accordingly, it is common for product design engineers to feel that they are in a more elite part of the organization than are the manufacturing engineers. At the same time, manufacturing engineers often feel that it is their group that deserves the most credit because without them the product probably could not be made.

A related source of problems arises when professionals who need to interact to get their own work done have goals that conflict with each other. In his study (Ginn, 1983) of three business units of a major chemical company, Ginn demonstrated this relationship to be significant at the R&D-production interface. The production system is mainly oriented to achieving optimal rates of production output, while R&D seeks to introduce new products. Because the commercial introduction of new products interferes, especially in the short run, with achieving maximum production output, a classic case of conflicting goals often exists. Ginn and Rubenstein (1986) report that the interdepartmental conflict between these groups and the corresponding effort to place (and avoid) blame were relatively high. A classic struggle of this sort is to argue, most likely through memo warfare, whether a design engineering group's major proposed engineering change order will improve a product's performance enough to justify a lengthy delay and sizable expense for the required retooling.

The workplace culture of engineers is likely to be a blend of the dominant values and norms of their profession coupled with cultural elements created through time by the management of the particular organizations in which these people work. In particular, formal reporting and review relationships, combined with criteria for the successful completion of a job, will strongly influence the type and intensity of "working together." Managers of engineers may attempt to encourage professional staff to work together, yet the scope and span of such collaboration will be limited by those individuals' backgrounds, education and training, as well as by any natural boundaries created by the organization of work in the company. Performance measures that focus on the optimization of the components of a system (a product or a manufacturing process), rather than on the system as a whole, is another common catalyst for specialization.

The legacy of engineering education

In this regard, blame has been cast more broadly at the education of engineers in the United States. The authors of a recent report, *Made in America,* summarizing the research of the MIT Commission on Industrial Productivity (Dertouzos et al., 1989), point to a trend in engineering curriculum. Since World War II, courses in engineering have turned away from issues of industrial practice with a corresponding increased emphasis on fundamental scientific and engineering principles. This trend, fueled by scientific developments that shaped new industries and government funding for military-based research at engineering schools, led to a retreat from the study and teaching of product development and process and production engineering.

The MIT Commission observes that this evolution "was both inevitable and desirable." The authors are critical, however, of the result: that faculty members lack industrial experience and that engineering graduates of top universities are trained for the world of research and development but not manufacturing or even for product design. As other universities tend to model their curricula on those of the leading programs elsewhere, this trend has become exacerbated. The study of computer engineering, for example, is not likely to include courses that cover topics such as the packaging of integrated circuits, their fabrication or assembly into larger circuits. This is not the kind of educational preparation that facilitates productive, collaborative industrial work at the design-manufacturing interface.

To make matters worse, there are very few university graduates in the United States who actually majored in manufacturing engineering. While many colleges are assembling such programs, until recently there were only a handful that existed. The exposure to manufacturing engineering provided by departments of mechanical engineering or systems engineering was not extensive. Much of what practicing manufacturing engineers learned in formal classroom settings was through technical schools outside the circle of formal, accredited undergraduate and graduate degree programs.

Finally, engineering education tends to be strictly technical in content and scope. It is generally lacking in emphasis on the development of written or oral communication skills. Furthermore, the high levels of competition that exist at the better undergraduate engineering schools breeds future professionals who lack practice at, and perhaps even a basic orientation towards, collaborative team projects.

The Culture of Product Engineering

What can we say about the workplace culture of the product design engineer? We address this question by piecing together findings from a fragmented literature and from our own recent experience. We intentionally look at several different kinds of industries, because product technology and competitive realities will shape the work of design engineers, both individually and collectively. To be effective, product design, which essentially consists of problem-defining and problem-solving activities, needs a managerial environment that guides and facilitates such work. Three key aspects of this culture, discussed below, are the handling of design deadlines, dealing with technological complexity, and forming effective design teams.

Traditional behavior around design deadlines

Despite its creative mood, product design needs to be managed. Mature companies often employ top-down decision-making to create environmental stability for those who do product design. Such an approach requires that deadlines for producing designs be set in advance and that those deadlines be taken very seriously. This emphasis on time-based deadlines creates a natural pressure for design engineers to attempt to modify existing designs without the kind of thorough analysis and testing that would normally be desirable. In their study of a large design engineering unit in a major U.S. automobile company, Liker and Hancock (1986) observed that one result of this approach was that extensive and unanticipated resources had to be allocated to the subsequent correction of problems. They also noted the vicious cycle that arose when future new products were neglected while scarce time and resources were applied to fighting fires from former faulty designs.

The important point here, from the perspective of managing the design-manufacturing interface, is that in some working environments, design engineers have become accustomed to meeting what they perceive as being serious though not critical project deadlines. The deadlines are serious enough that one does not wish to fail to have a product design ready at the pre-established time. Apparently, organizationally imposed criteria—the "timely hand-over" of a design—can dominate the professional norm of producing a high-quality design. What makes this compromise possible for the design engineer are two other elements of the work culture: it is acceptable, and even expected, that change orders be issued to "fine tune" the design after its release; and that it is usually someone else's responsibility at that point.

This approach to product development can be functional for the design engineer in that there is a built-in opportunity to "look good under pressure" on two occasions: first, when the design is released, and then again when the fine tuning is accomplished. To the extent that design engineers are rewarded for such behavior, and the rest of the organization finds it to be acceptable practice, these engineers are not likely to instigate a radical change in the process of new product introduction. Nor is a top management statement about the importance of doing product innovation right, in the absence of any major structural changes such as introducing cross-functional teams or new performance measures, likely to alter the traditional pattern of behavior.

Companies that would like to achieve "mature first costs"—where the new product is designed so that it can be manufactured as intended from the beginning and at the target cost—must take appropriate steps to overcome the culture that naturally resists this objective (see Note 3 at end of chapter). This kind of planning and scheduling can be particularly difficult to accomplish where there is extensive specialization in the design of a complex product. Here, the aggregation of component designs into system designs is an independent activity and the release of the full product design must be carefully monitored and controlled.

Technological complexity and the culture of design

The technological challenges facing design engineers, given their base of knowledge, vary greatly from industry to industry and from company to company. Each design engineer has a personal repertoire based on prior experience in defining and solving particular technical problems. In each company engineering solutions are in part constrained by the scope and content of existing data bases. One should not be surprised to find that the extent of technological complexity inherent in traditionally designing a certain class of product in part determines the culture of that organization's design function.

Consider the following illustration. In large complex organizations with a relatively stable and well-understood product line, engineers often work in a hierarchical structure. Supported by experienced draftsmen, and engineers in supplier firms (and consultants when needed), those with the title "design engineer" may not require extensive technical knowledge. Liker and Hancock (1986) found that such engineers built experience mostly in "coordination of information flows and policing engineering changes," rather than in developing new, effective design solutions. These researchers observed that the culture of such an organization is likely to resist the development of specialized technical competence. Instead, in these kinds of companies, ambitious young engineers are encouraged to concentrate instead on behavior that will lead to a rapid promotion out of product engineering and into management.

The same study also found that smaller companies tended to produce state-of-the-art products and build a strong record of success in satisfying the requirements of their market niche. These companies traditionally have other distinguishing features in their product design culture. They are likely to be engineering-driven, as in the example of Company C (from the Introduction to this chapter). Here, design engineers often have the power to imagine the next technological breakthrough and tell marketing managers what they will provide and approximately when this new product will be available. It is not unusual, in such settings, for design engineers to incur little or no penalty for missing their deadlines. Designers and their managers in such settings are likely to be sympathetic to the notion that product design is a creative process which is inherently hard to forecast.

Technological complexity may also be a catalyst for developing an effective design–manufacturing interface if the company has made a concerted effort to balance the power among design engineers, manufacturing engineers and marketing experts. Company A (again, from the Introduction) was founded to build a complex state-of-the-art computer product. They formulated a philosophy, and worked hard to sustain it, whereby technological complexity became an incentive to collaborate. For their design engineers, getting the product successfully built on time was the single preoccupation, rather than winning cross-functional battles or rising rapidly to the managerial ranks. Difficulties in achieving such collaboration among the key technical actors at the design-manufacturing interface are discussed below, in the concluding section of this chapter.

Forming effective design teams

There is considerable middle ground between the two extremes of routinized, hierarchical design and free-flowing creative design engineering. Here one finds room for effective product design that, in timing, cost, and quality, fits a particular window of market opportunity. Such product design efforts need to be conducted with a combination of firmness and flexibility.

The high technology world of computer design (alluded to in the Company A example) has supplied us with perhaps the most explicit published description of this type of product engineering and its management: Tracy Kidder's book called *The Soul of a New Machine*. This book provides a vivid sense of what it was like to be on the team of young engineers and software specialists working to build the new Eagle computer for Data General in 1979. This particular computer design project has now become a classic in the literature of technology management. It captures the enthusiasm, pressures, managerial styles, and team spirit of a high visibility "bet-the-company" new product development effort. In other companies with similar competitive pressures, technological challenges, and professional resources, the dominant features of the product engineering culture might well be the same, even if the personalities of the team members and project manager are different.

Kidder places little emphasis on the professional background of the individual designers. What emerges as being significant is the process of self-selection through which individual designers having high-native intelligence, top quality technical training, and a willingness to work extremely hard come to work together. Kidder points to "a mysterious rite of initiation," called "signing up," in which members of the design team, from the initiation of their involvement, agree to place the success of the project ahead of all personal and family concerns. The joy from anticipating the success of the project, doing something that had not been accomplished before under intense time pressure—in short, the opportunity to "get a machine out the door with their names on it"—was something that they each valued highly.

The key elements of culture, according to Kidder, were traceable to the informal process of the design team and the particular style of the project manager. Particularly notable, technical solutions could not be traced to individual designers, due to the informal, interactive, and rapid progress on particular issues. Everyone seemed to understand that the team members were there to do high-quality innovative design and little else; the project secretary voluntarily served as a buffer between the team and the company's bureaucracy, thereby shielding the young designers from unnecessary distractions.

The project manager began by setting an optimistic product development schedule: "the earliest date by which you can't prove you won't be finished." Then he served as a cheerleader with a particularly aggressive stance. To the rest of the company, he predicted great progress, thereby enhancing the pressure felt by his designers. He trusted his professionals to accomplish broad tasks. Though he established and enforced the rules for the design of the new computer, he did not try to tell each of his team members exactly what to do. Meanwhile, the president of the company encouraged competition among product design teams, allocating resources incrementally to the team whose ideas for new products seemed most exciting and achievable.

Tracy Kidder captured the life-or-death quality of new product development in the highly competitive and dynamic computer industry. The intensity of the product design effort mirrored the chaotic industry environment which their company faced. The company's youth was echoed by the youth of the product designers. Nowhere was there room for traditional, cautious approaches. Product engineering in this case, and in many others like it, is an elite, high-recognition, and unforgiving work environment. Whether this particular approach was totally effective at Data General in 1979, or more important for our purposes, whether it is likely to be effective at a different time or in a different company, is questionable.

Another example of the culture of product engineering differs from the others outlined above. However, it shares the theme that product design tends to be an identifiable professional activity with a culture that must fit that of the company as a whole. The composition and structure of an effective design team, then, may vary from one industry and technology to another. Nevertheless, in all cases one must ascertain its fit with the world of manufacturing.

The Culture of Manufacturing Engineering

The working world of the manufacturing engineer probably is documented even less than that of the design engineer. One reason for this is that potential audiences for the business or trade press are not likely to consider it exciting. Even when accounts of manufacturing success (or failure) are made, emphasis is usually placed on commercial or technological performance, rather than on the underlying engineering activity. Furthermore, the job category of manufacturing engineer is less well-defined than its counterpart positions in product design. In this section we look at conventional and more futuristic notions of manufacturing engineering. Special attention is given to the impact of company and plant size and trends toward more scientific images of manufacturing.

A conventional view of manufacturing engineering

In a recent text on manufacturing engineering, Daniel Koenig (1987) divides the field into four distinct functions: advanced manufacturing engineering; process control; methods, planning, and work measurements; and maintenance. Advanced manufacturing engineering is the function of direct interest in this chapter because it is naturally connected more closely to product design than the other functions, which concentrate on operational aspects of satisfying manufacturing requirements.

As outlined by Koenig, advanced manufacturing engineering typically contains the following functions: area planning, capacity analysis, capability evaluations, new technology evaluations and needs, producibility engineering, computer-aided manufacturing development, investment project management, and long-range planning and forecasts (see Note 4 at end of chapter). After identifying these functions, Koenig goes on to say:*

> A separate engineer could have each or several of these responsibilities, or several engineers could share one area of responsibility. It matters little what the specific organizational structure looks like as long as all responsibilities are properly attended to.

However, for our purposes, it *does* matter to some extent what the specific organizational structure looks like and what the scope of job

---
*Koenig, p. 6.

responsibilities are. For example, if there are no specialists formally assigned to investigate the manufacturability of proposed new designs, then the culture of manufacturing engineering is oriented primarily to the improvement of on-going production. To the extent that manufacturing engineers, under this scenario, get involved in matters of new product introduction, it is likely to be a "sideline" for them. Their workplace culture will emphasize the maximization of output, fixing problems on the factory floor, and planning future production processes, given demand forecasts and product designs developed by others elsewhere in the company. These cultural imperatives clearly are in conflict with effective new product development.

Producibility engineers and design reviews

In contrast with the above scenario, consider the situation where a company has organized manufacturing engineering so that it includes an independent group of producibility engineers. According to Koenig, producibility engineers are charged with working along with design engineers:

> A producibility engineer...uses as inputs designs from design engineering and capabilities observed in the factory, then converts those designs and capabilities into workable designs so they can be made in the factory. The producibility engineer, then, is a compiler of information, an optimizer of factory input and design input into a producible scheme.*

In practice, the primary function of the producibility engineer, then, is to initiate formal design reviews at the point when a design engineer is ready to present drawings or layouts of a part or product assembly. The review session contains an exchange of information between the design engineer and the producibility engineer, whereby the former explains the logic and motivation for the particular proposed product design and the latter explains either the complications that this may cause for manufacturing, or the existing manufacturing capabilities that are not being adequately called for in the design. Written memoranda document such review sessions, thereby setting the agenda for subsequent problem resolution.

The workplace culture of the producibility engineer is generally in the spirit of successful new product development, but with particular built-in conflicts. The producibility engineer's role is akin to that of an ambassador seeking adequate representation for his homeland (manufacturing) among those of foreign soil (product design). To be effective, such ambassadors must be excellent communicators. They are

---

*From Koenig, p. 113.

likely to be judged on their ability to stop the equivalent of a foreign invasion. This type of manufacturing engineering, aimed directly at the design-manufacturing interface, is a necessary step in a traditional, functionally-dominated organizational structure. It is the manufacturing organization's attempt to gain parity in the "chimneys of power," whereby problems, once surfaced, are pushed to higher levels for ultimate resolution.

While Koenig explains the tensions between the orientations of the producibility engineer and the design engineer, his conclusion lacks a note of reality:

> In cases where there is disagreement between the producibility engineer and the design engineer, the procedure is to determine the correct amount of design safety margin that is best for the overall interest of the business.*

Clearly, this author has written about how the world should work, rather than how it does work (see Note 5 at end of chapter). As discussed later, the notion of joint ownership for new product introduction, central to most current models of organizational effectiveness, is different from what transpires through the formal use of a producibility review by manufacturing engineering.

The impact of company and plant size

Manufacturing people generally seem comfortable with the following prescription for their work: take instructions, work from a complete design, and then get things done! The traditional hand-off from design to manufacturing is like the beginning of the last leg of a relay race. The manufacturing team is the last runner in providing a product that can be introduced to the market and will try to make up ground in the sense of figuring out how to produce the product as currently designed, how to adjust manufacturing processes in response to product design changes, and how to ramp-up, as rapidly as possible, to commercial rates of production. Establishing a workplace culture that supports such efforts is a challenge unto itself.

To some extent the size of the company, as reflected in the scale of operations, is likely to affect its manufacturing culture. Company size brings other important variations, including the extent of resources and support for on-going professional development through company-sponsored training programs or tuition reimbursement plans for employees pursuing advanced degrees. Additional factors that tend to vary with size are the structure of the manufacturing organization,

---

*From Koenig, p. 135.

the number and type of manufacturing engineers, and the locus of power between corporate staff and plant-level personnel.

In large companies, manufacturing engineers may work in groups of similar specialists where narrow technical competence is valued and rewarded. Such specialization brings the possibility of more sophisticated solutions where manufacturing engineering problems are naturally decomposable. However, it also brings the risk of unproductive lapses in communication and coordination when technical problems are more interrelated. The existence of independent groups of specialists also tends to generate turf battles, particularly over issues of resource allocation.

Advanced manufacturing groups are also common at the corporate level in larger companies. Members of such groups often have advanced degrees and relatively high salaries, thereby tending to set them apart from manufacturing engineers in individual plants. The corporate groups tend to be especially familiar with the latest computer-based automation technologies, although they may lack current, in-depth familiarity with the manufacturing capabilities of particular plants. Typically, they function as internal consultants to the plants in times of modernization, capacity expansion or major new product introduction. In such instances, bridging the cultures of the corporate and local manufacturing engineering groups can be a challenge unto itself. Larger companies with this structure need to leverage the investment they have made in advanced skills at their corporate groups. Plant-level groups may resent that this investment was made in the first place. Stakes can be high for the company in terms of both return on investment and lost opportunity cost if such efforts are not fruitful.

Smaller companies, in contrast, tend to have very few highly trained and experienced manufacturing engineers. They are more likely to employ generalists at the plant level and to operate without a corporate advanced manufacturing group. Manufacturing engineers (whatever title they work by) tend to develop experience with particular sets of machinery, fixtures, tooling or software. Frequently, personnel turnover in these positions is not high and much of the institutional memory about the manufacturing capability is in minds of individual engineers. Documentation of prior projects may be spotty. Those who worked on some component of the plant may informally inherit the responsibility for all subsequent problems or modifications. In such a setting the mood is likely to be more pragmatic and less bureaucratic.

Current demographics and future trends

To get additional views of the culture of the manufacturing engineer in the United States, we must turn elsewhere. A reasonable place to start is with some demographics. A recent survey by Koska and Romano (1988) of the consulting firm A. T. Kearney, sponsored by The Society of Manufacturing Engineers (SME), looked at the future of manufacturing engineering in the United States. More than 7,500 manufacturing practitioners participated in a questionnaire exploring their views on manufacturing engineering today and requirements for the future. This survey is of interest to us in terms of both its participants and their responses.

A summary of the survey sample provides a glimpse of the demographics of this population of manufacturing professionals in the United States. Of the respondents 98 percent were male with an average age of 44 years. Compared with other professional fields, the level of education among these manufacturing practitioners was not high: 55 percent of the respondents completed at least a four-year bachelor's degree and 26 percent had other formal training. These figures, for our purposes, must be put into proper perspective. For example, 27 percent of the respondents classified their current job as being "manufacturing engineering" (but this fraction would roughly double if one included other job titles with engineering labels such as tooling, systems, process, production, N/C (numerical control) analyst, mechanical and industrial), while 28 percent were managers (probably including a large number of former manufacturing engineers) and the remainder, less than 20 percent were in other technical specialties. Compared with their counterparts in product design, manufacturing engineers are probably older and have less formal education (see Note 6 at end of chapter). Naturally, this would not be true for all companies, and in any given company there would be some counterexamples, but it is an important generalization to keep in mind when comparing the cultures in the two professional groups.

Some of the quoted comments presented in the report on this SME survey emphasized the personality and dominant work style of manufacturing engineers. The need to be more creative and imaginative was stressed. Enhanced communication skills ("I never understand what these people are talking about") was highlighted. The requirement for manufacturing engineers to be able to understand manufacturing problems in the context of broader business issues was also perceived.

Such findings from this survey clearly challenge those who educate and train manufacturing engineers to supplement traditional technical subjects with a wider range of problem-solving and social skills. In the words of one respondent, it is no longer sufficient that a manufacturing engineer be "a genius with numbers or computers." One cannot help wondering, however, how much improvement on such matters is reasonable to expect. Admittedly, undergraduate engineering students are at an age where their professional orientations and sense of needed skills are quite formative. Nevertheless, considerable self-selection takes place among those who have chosen to become manufacturing engineers, and also among those who educate them and set their curriculum. No doubt substantial progress in these directions is possible, but it would be naive to predict radical, rapid change in personal and problem-solving styles across the broad span of U.S. manufacturing engineers.

A different scenario for manufacturing engineering is, however, possible. When one looks to the leading manufacturing companies in other countries, notably in Japan and Germany, the demographics and workplace culture of manufacturing engineers stand in stark contrast to the typical U.S. situation as presented. It is not unusual in such companies to find Ph.D. engineers on the plant floor. Moreover, in the Japanese computer industry it is common for top engineering graduates to be reassigned between product design and manufacturing (Westney, 1986). While there are no statistics comparable to those from the United States cited above, the work of J. Jaikumar (1986) suggests that such radical changes in the role of engineers in manufacturing are both possible and necessary. Consider, for example the case of Hitachi Seiki, a large Japanese machine tool manufacturer. In 1980 this company established an Engineering Administration Department, consolidating the functions of machine design, software engineering, and tool design and stressing the more generic role of systems engineer. The head of engineering performed this reorganization to enhance the coordination among these functions.

The culture and skill base at Hitachi Seiki evidently supported this plan. A team of just 16 engineers simultaneously designed, built, and successfully installed 3 different flexible manufacturing systems within an 18-month period. This tight-knit group of systems engineers completed these new product introductions (for their own internal use) in record time and consistent with the company's preestablished return on investment criteria. In summary, manufacturing engineering cultures in technology-oriented U.S. companies may shift over time to be more like those at Hitachi Seiki. To the extent that this shift is widespread, manufacturing engineering will look very different from the conventional views of authors like Koenig or the more incremental extrapolations of SME members. Despite this glimpse of the future, most U.S. manufacturers today are seriously constrained by the demographics and organizational inertia outlined above.

Implications of more science in manufacturing

The world of manufacturing is experiencing dramatic technological shifts with struggles to adopt and adapt new forms of computer-based automation becoming commonplace. As was the case at Hitachi Seiki, a new breed of manufacturing engineer, highly computer-literate, is accompanying the trend towards flexible manufacturing technologies. The worldwide community of manufacturing engineering now speaks freely of the likelihood that we are approaching the age in which "manufacturing science" is an apt label rather than a contradiction in terms.

The recent influx of computer-based advanced manufacturing technologies—such as CAD/CAM, complex computer simulations and CIM data bases and software systems—has escalated a cultural conflict between artisan and engineer. Artisans are experience driven. They tend to draw heavily on tried and true, often basic, modes of problem identification and resolution. Modern engineers, in contrast, are comfortable with using the latest technological tools in conducting their work. Observers have noted that this conflict between artisan and engineer has characterized American manufacturing practice for several decades. It is hard to imagine a manufacturing "science" that does not encapsulate the vast experience that has accrued in the use of basic production processes and materials. Cultural tensions of this sort will affect an organization's ability to achieve the kinds of enhanced design-manufacturing integration that new technologies facilitate.

Some observers see the increasing scientific orientation in manufacturing as a sign of progress. Others, however, including Kim Clark and Robert Hayes (Clark, 1988), are concerned that the purely scientific view contains an unacceptable down-side risk, that is, a static view of the world of manufacturing. To build more effective production capabilities, the argument goes, manufacturing professionals need to emphasize human problem-solving skills and a dynamic perspective consistent with the changing conditions likely to dominate in the future world of manufacturing. So doing, given the current status and culture of this field, will not be easy.

Perspectives on Managing the Design-Manufacturing Interface

Organizing for manufacturable design is a specific example of a more general challenge: improving the coordination between groups that seem to have developed a largely independent, and in some respects, win-lose relationship with each other. As Chester Barnard (1938) pointed out long ago, people contribute their individual activities to the organization that employs them, but it is the collection of those internal activities, coupled with transactions with the external environment, that will determine the organization's success. Internal activities include the operations of research, development, and manufacturing, the staff efforts that support operations and special activities that span the traditional boundaries within an organization. Transactions with the external environment include such activities as R&D outreach, purchasing and supplier management, sales and market development, and after-sales service. According to Barnard, it is the function of the executive to create a setting in which all of these individual actions become effective in the pursuit of organizational goals.

How, then, might the work of product designers and manufacturing engineers be better integrated to produce quality products faster and cheaper? The literature cited above argues persuasively that one must go beyond normal managerial hierarchies to develop tighter integration through devices such as individual coordinators or cross-functional teams. In their book on *Dynamic Manufacturing,* Hayes, Wheelright, and Clark (1988) advocate particular approaches. They see innovative forms of project management and more interactive information exchange through "overlapping problem-solving between upstream and downstream groups" of specialists as providing a more effective and less formal mode of integration. Such initiatives, described in more detail by Clark and Fujimoto (1987), forces specialists to work with and learn from other specialists, and gradually tend to become more like generalists themselves.

Along similar lines, the recent forecast of the role of the U.S. manufacturing engineer in the twenty-first century projects that a central activity will be serving as "operations integrator," as distinguished from "technical specialist." This role will require broader communications skills and more sophisticated technical skills than is typically present among manufacturing engineers in the United States today.

Such changes in organizational structure or job design raise other simultaneous challenges. For example, Ginn and Rubenstein (1986) persuasively argue that fundamental differences in goal orientations at the R&D/production interface may thwart the efforts of such integrators. They conclude that direct effort to identify and remove goal incompatibility has to be a high priority in any action plan. Critical to the success of such a plan is the creation of proper incentives for all to work, in a genuine team setting, on the successful introduction of new products, and the ability to measure performance in terms of these new incentives (see Note 7 at end of chapter). It is unrealistic, however, to expect widespread acceptance for unfamiliar and, for many people, unnatural changes in workplace culture. We now turn to some classic concepts in the management literature as a framework for suggesting necessary actions and inherent challenges.

Blending cultures through organic management structures

What seems to be missing in many current investigations of problems at the design-manufacturing interface is a sense of the subtlety that anthropologists bring to the study of culture. Anthropologists, for example, have a healthy respect for the challenges and risks in attempting to blend cultures or to create new ones. Burns and Stalker (1961), in their pioneering book, *The Management of Innovation*, adopted this perspective and their thoughts seem very appropriate at this point.

Burns and Stalker, for example, observe that designers are comfortable dealing with rough sketches, while manufacturers need precise drawings. Designers are accustomed to improving their designs even after they are "complete," while manufacturers prefer stability so that they can perfect their production methods. Managers who want to encourage these kinds of professionals to work together more intensively must not grow impatient when such innate differences in style lead to frustration on the parts of the various participants.

Burns and Stalker also consider fundamental challenges of communication. For example, specialties (such as design and manufacturing) naturally create their own language and words come to embody special meanings. Thus even if representatives from the two cultures are brought together in a new working relationship, one must expect a certain amount of confusion due to inherent difficulties (rather than resistances) in communication. Cross-training, either in formal educational programs or on the job, can alleviate some of these language difficulties. To some extent, however, the problem is more subtle because some of the confusions will be due to tacit understandings built up through years of working on particular problems with special points of view.

Burns and Stalker make a lasting contribution in their conceptual description of "mechanistic" and "organic" systems of management. The mechanistic system is characterized by specialized jobs, controlled supervision through hierarchical structures, and extensive vertical interactions (between superiors and subordinates). As outlined earlier in this chapter, much traditional product design and manufacturing engineering in large companies occurs under this type of management system. Burns and Stalker argue that such an approach is appropriate under stable conditions, which in our case could refer to factors such as market requirements, technological core, and relative change in competitive behavior.

The "organic" form, these authors argue, is the more appropriate system of management under the changing conditions that characterize much of today's world of manufacturing. Under this form of management, individuals contribute their special knowledge and experience to the common task of new product introduction. A community of interest exists, with shared commitment to the joint enterprise. Most communication is lateral rather than vertical, and the content of information is largely the sharing of information and advice, not just instructions and decisions. Most important, perhaps, is that individual tasks be adjusted continually and redefined through interaction with others.

The quest for specialization and integration

A fundamental managerial challenge is embedded in such structural initiatives. The pioneering work of Paul Lawrence and Jay Lorsch (1967) presents the organizational dilemma of specialization in group problem-solving contexts. Their differentiation-and-integration model of management applies well to issues at the design-manufacturing interface. Following the prescriptions of this model, participants in new product introduction must work in highly "differentiated" settings to the extent that they are dealing with different kinds and levels of uncertainty in the work that they must perform.

Such needed differences, Lawrence and Lorsch discovered, sometimes involve fundamentally different ways of thinking and behaving. Hence, it is not surprising that product design engineers and their traditional work setting might be different from their counterparts in manufacturing. Designers are oriented to innovation. They try to develop new products either in response to their sense of market needs or, in some engineering-driven companies, simply to test their ability to create new things. Manufacturing engineers, in contrast, traditionally have been concerned with the achievement of efficiencies in the use of available resources to make current and new products. To the extent that, at any one point, there is a high degree of natural differentiation between such two groups, the Lawrence and Lorsch model postulates that it will be difficult to achieve integration between them.

Another critical question asks under what conditions will such pursuit of structural change be fruitful? In a more recent book, *Renewing American Industry,* Lawrence and Dyer (1983) speak to this question. These authors develop a theory of "the readaptive process," which emphasizes the forces that favor or hinder an organization's quest for improved performance. This theory can be seen as an extension of the earlier work of Lawrence and Lorsch in that it also concentrates on the inherent tensions in achieving the ability to pursue innovation and efficiency at the same time. The readaptive process, as its name suggests, is a dynamic theory of change that is meant to apply across industries. The authors look at external forces including the development of core technology, public policy, and inter-industry and foreign competition, as well as internal aspects of organizational choice. Key variables—affected by these forces, and in turn influencing adaptation—are information complexity and resource scarcity.

For our purposes the forms and means of organizational adaptation are the most interesting aspect of this theory. The particular form called "readaptation" may thrive when there are intermediate levels of both information complexity and resource scarcity. This creates a state of environmental pressure for a company to "learn and strive" to readapt—with too much information complexity innovation will be restricted and with too many or few resources the quest for efficiency is destined to fail. Under such a balanced set of conditions, organizations are most likely to achieve a combination of high differentiation and high integration. Power, then, will be reasonably shared among managers and professionals. High levels of efficiency, innovation, and employee involvement can be achieved. In this situation readaptation, a critical factor, is facilitated through three distinct means to motivate employees: internal market incentives (profit centers, bonus plans), bureaucratic procedures (hierarchical command and control), and clan-like features (strong, interactive working relationships). In this theory of readaptation, Lawrence and Dyer have offered helpful yet general perspectives about organizations and their environments that can be applied to specific internal organizational activity such as the design-manufacturing interface.

Closing the culture gap

T. G. Whiston (1988) recently prepared a monograph dedicated largely to the proposition that companies must get better at developing and executing organic forms of management. This monograph, which raises many good questions about the nature of integration and associated forms of information processing, outlines the general challenges of organic structures. Although it touches on matters of product design and manufacturing, it does not specifically deal with cultural aspects of effective work at this interface. The key managerial question, then, is how to bring about the transition to an organic structure for new product introductions?

The field of socio-technical systems (STS), which is described in a classic monograph by one of its founders, Eric Trist (1981), approaches work design with a commitment to achieving organic management structures. James Taylor et. al. (1986) report the case of Zilog, Inc., a company that successfully modified its Component Design Engineering (or CDE) organization, using socio-technical systems (STS) approaches, to operate in a more organic fashion. This case involves product design personnel but not manufacturing. Nevertheless, it is relevant to us for its careful description of an attempt to create a new workplace culture for engineers.

The case study covers the shift at Zilog from small projects to design 8- and 16-bit microprocessors to more complex projects to design 32-bit processors in the early 1980s. CDE had accomplished its earlier microprocessor designs through a traditional functional organization, containing separate groups for architecture, logic and circuit design, layout design, and several other activities. Despite conflict and friction among these groups, this structure seemed to work for the small chip development projects that required only one architect, one design engineer, and one layout designer.

Under this traditional form of organization, the layout supervisor served as an intermediary between the design engineers ("an elite caste with individual star performers") and the layout designers ("a clannish underclass" who "often feel like second-class citizens"). The layout specialists used a recently adopted CAD system to produce graphic renditions of circuits created by the design engineers. Since layout assignments were made daily, the layout designers had no ongoing commitment to any particular project. Furthermore, they had trouble with the new CAD technology which had unanticipated shortcomings. Design engineers were dissatisfied with the attitude of layout designers and with their ability to meet work schedules. Turnover of layout specialists was high.

The director of CDE, being sensitive to the growing problem, decided to use STS analysis to facilitate a more effective organizational design. A design team, including representatives from all functions and levels at CDE was formed. One of the benefits of this STS analysis was the realization that Zilog had several different views of itself as a manufacturing company, a marketing company, and a design company. Top management was able to use this information to develop a single mission statement and philosophy, thereby taking a major step toward integrating the several operating functions of the company. The STS team, in turn, decided that CDE ought to work more closely with marketing and manufacturing. Finally, a new CDE organization was formed, in which layout designers and engineers worked together on one of four permanent, integrated project teams. New work space and flextime arrangements facilitated effective project performance. New performance measures, aimed at tracking project progress and outcomes, strengthened incentives for teamwork. Problem-solving and decision-making was much more interactive. Finger pointing across specialties, in response to project setbacks, ceased.

More than a year after the new CDE organization was implemented, an informal assessment disclosed almost uniformly higher morale and increased awareness of the total design process. Interestingly, however, the following negative reactions were also voiced:

- Managers' frustrations over the longer time required for decisions by consensus (rather than by "benevolent dictatorship")
- Engineers' concern that consensus does not always yield the best technical decision and that too much time seemed to be spent in meetings
- A general feeling that professional relationships within specialties were diminished as a result of the product-team structure.

Even success stories like Zilog raise concern that there is more at stake in trying to blend cultures than readily meets the eye. While we may have identified the types of organizational structures that are consistent with improved integration between design and manufacturing, the literature says little about the set of actions that will be sufficient. For example, these new forms of cross-functional project teams with operations integrators will call for unusual methods for resolving conflicts and compatible human resource policies.

In these innovative forms of organizing for manufacturable design, no longer does one engineering specialty automatically get veto power over the others; new modes of consensus-building at cross-professional working levels are required. No longer do issues routinely get escalated to top management where the outcome will be based largely on familiar power struggles across hard functional boundaries. To successfully resist this "natural" bureaucratic tendency calls for new types of leaders. Leaders must have an ability to develop staff to make as many decisions as possible, rather than to refer all issues for higher review at the slightest sign of possible difficulties. At a more personal level, such leaders will have to learn to resist the tendency to eagerly grasp onto emerging disagreements as opportunities to demonstrate their own decisiveness and power-brokering abilities.

Programs to hire and train engineers will have to reflect the special social demands of these new jobs. There is also the clear danger of professional obsolescence among those who do not learn to operate in such new modes. Among those engineers who have worked for many years in established and traditional job settings, such adaptations are likely to result in considerable stress. In extreme cases incidents of "burnout" or mid-career crisis will occur. Regardless of particular workplace cultures, it will probably be necessary for companies to reassess their hiring and human resource development practices. Shapero (1985) makes this point with reference to professional work in general and it is particularly true for the type of rapidly evolving engineering jobs that are of concern in this chapter.

Summary

The Zilog experience of being forced by external pressures to seek fundamental internal change was probably a precursor of what is becoming a major trend among U.S. manufacturers. Regardless of whether socio-technical systems (STS) or some other approach is used to design and implement fundamental change in workplace structure, many companies, such as the three that were referenced at the beginning of this chapter, are taking positive steps to improve the design-manufacturing interface. Leading-edge companies are moving their new product development processes further away from the mechanistic end of the Burns and Stalker continuum and closer to the organic end. For many mature companies the forces favoring readaptation may be lost unless proper action is taken soon. Implementing such change remains a primary challenge for senior management today, particularly in organizations that grew prosperous through the effective use of the mechanistic model when conditions of stability prevailed.

The challenge can be summarized in terms of several fundamental tensions that were described in this chapter:

Achieving strong collaborative ("clan-like") behavior among specialists with different backgrounds. The goal is to create a daily belief that internal collaboration is better than either competition or independent problem solving. The challenges include: achieving an effective blend between science and experience; making technological complexity a catalyst to achieving integration, rather than being a formidable barrier; and establishing a common vision that promotes new productive modes of conflict resolution.

Balancing firmness and flexibility in bureaucratic procedures. The goal is to provide a culture which supports simultaneous attention to innovation and efficiency. The challenges include instituting design deadlines without overly constraining engineers from doing their best work, encouraging the introduction of new technology without missing strategic market windows, and developing project managers who can combine the skills of leaders and cheerleaders.

Providing a positive internal market mechanism. The goal is to establish compatible objectives for the various participants in new product development. The challenges include establishing more appropriate performance measures, creating team-based incentives for professionals used to independent assessments of their work, and encouraging participants to help create new work settings that differ radically from those to which most have already become comfortable and proficient.

Based on the evidence to date, one cannot assume that dealing with these tensions will be easy. Patience and perseverance almost certainly will be required, for there is no easy "cookbook solution" to fit all situations. Companies that manage to achieve one successful new product introduction under an experiment with workplace culture still need to become skilled in institutionalizing such successes. Without true cultural reform, the re-emergence of familiar gaps can be expected.

Notes

1. Design engineers and manufacturing engineers are only two of the key players in the introduction of new products. Other specialists who may play a part in such efforts include those from marketing, industrial design, test engineering, quality assurance, purchasing, operations, certain vendors, and field service. This chapter deals with the limited topic of bridging the cultures of design and manufacturing engineering rather than the more comprehensive version that would include all of the key players.

2. From the viewpoint of research, then, this chapter hopes to point to areas that merit more systematic empirical investigation. Its primary purpose, however, is to provide preliminary insights for practitioners who wish, in effect, to change the engineering cultures in their organization to promote more integrated design and manufacturing.

3. The current project of the Boston University Manufacturing Roundtable, referenced in the opening footnote of this chapter, has this research objective.

4. Describing each of these manufacturing engineering responsibilities is beyond the scope of this chapter, as is coverage of the various specialties within the job category of design engineer. Accordingly, there is little attempt to explain the substantive work content of someone with either of these two titles. We concentrate instead on aspects of the underlying work culture that is likely to be relevant in attempts to improve the integration of their work.

5. The subtitle of Koenig's (1987) book is "Principles for Optimization." This provides a clear signal that he is not trying to explain the cultural realities of being a manufacturing engineer today. There is nothing wrong with adopting a normative focus of this sort. Unfortunately, an excessive preoccu-

pation with the principles of being an outstanding manufacturing engineer can easily lead to a lack of sophistication about the types of organizational structures and cultures where such principles readily can be implemented. Koenig has adopted a more traditional set of organizational boundaries and their associated constraints, while this chapter (and the book of which it is a part) seek the creative destruction of rigid, functionally oriented forms of new product introduction.

6. One has little choice but to be speculative on this issue of professional demographics because we are dealing with two hybrid categories. Neither "manufacturing engineer" nor "product design engineer" is a traditional category for gathering national employment statistics. The National Academy of Engineering collects its statistics based on six traditional disciplinary fields of engineering in which almost all engineering degrees are granted, with a final category called "other"; there is no way to transform such statistics into their "manufacturing" and "product design" components. Bureau of Labor Statistics surveys might come closer to the mark in terms of employment levels because they are oriented to functional work activities. However, this potential data source would lack the demographic detail of interest to us in this paper.

7. The need to focus on broader organizational goals rather than more localized, functional goals, and to develop performance measures that encourage such behavior, is an increasingly common theme in the literature. See, for example, the OPT principle of manufacturing described by Goldratt and Cox (1984), in their novel, *The Goal*, and the work by Dixon, Nanni, and Vollmann (1989) on performance measurement in manufacturing organizations.

References

Badawy, M. K. "Understanding the Role Orientation of Scientists and Engineers," *Personnel Journal*, June 1971, pp. 449–485.

Barnard, C. I., *The Functions of the Executive*, Harvard University Press, Cambridge, Mass., 1938.

Barney, Jay B. "Organizational Culture: Can It Be a Source of Sustained Advantage?," *Academy of Management Review*, vol. 11, no. 3, 1986, pp. 656–665.

Burns, Tom and G. M. Stalker, *The Management of Innovation*, Tavistock Publications, London, 1961.

Business Week, Special Issue on "Innovation in America," 1989.

Clark, Kim B. and Takahiro Fujimoto, "Overlapping Problem Solving in Product Development," Harvard Business School Working Paper, 1987.

Clark, Kim B. and Robert H. Hayes, "Recapturing America's Manufacturing Heritage," *California Management Review*, Summer 1988, pp. 9–33.

Davis, Stanley M., *Managing Corporate Culture*, Ballinger, Cambridge, Mass., 1984.

Dean, James W., Jr. and Gerald I. Susman, "Organizing for Manufacturable Design," *Harvard Business Review*, vol. 67, no. 1, Jan.–Feb. 1989, pp. 28–36.

Dertouzos, Michael L., Richard K. Lester, and Robert M. Solow, *Made in America: Regaining the Productive Edge*, The MIT Press, Cambridge, Mass., 1989.

Dixon, J. Robb, Alfred J. Nanni, and Thomas E. Vollmann, *Breaking the Barriers: Measuring Performance for World Class Operations*, Dow Jones–Irwin, Homewood, Ill., 1989.

Ginn, Martin E., *Key Organizational and Performance Factors Relating to the R&D/Production Interface*, Ph.D. dissertation, Northwestern University, Dept. of I.E.M.S., Evanston, Ill., 1983.

Ginn, Martin E., and Albert H. Rubenstein, "The R&D/Production Interface: A Case of New Product Commercialization," *Journal of Production Innovation Management,* vol. 3, 1986, pp. 158–170.
Goldratt, Eliyahu M. and Jeff Cox, *The Goal,* North River Press, Croton-on-Hudson, N.Y., 1984.
Hayes, Robert H., Steven C. Wheelwright, and Kim B. Clark, "Managing Product and Process Development Projects," *Dynamic Manufacturing,* The Free Press, New York, 1988, pp.304–339.
Jaikumar, R., "Post-Industrial Manufacturing," *Harvard Business Review,* November–December 1986.
——, "Hitachi Seiki (A)," Harvard Business School Case, 9-686-104, 1986.
Kidder, Tracy, *The Soul of a New Machine,* Atlantic–Little, Brown, Boston, 1981.
Koenig, Daniel T., *Manufacturing Engineering,* Hemisphere Publishing Corp., New York, 1987.
Koska, D. K. and J. D. Romano, *Countdown to the Future: The Manufacturing Engineer in the 21st Century,* Society of Manufacturing Engineers, Dearborn, Mich., 1988.
Lawrence, Paul R., and Davis Dyer, *Renewing American Industry,* The Free Press, New York, 1983.
Lawrence, P. R. and J. W. Lorsch, *Organization and Environment: Managing Differentiation and Integration,* Division of Research Graduate School of Business Administration, Harvard University, Boston, Mass, 1967.
Liker, J. K. and W. M. Hancock, "Organizational Systems Barriers to Engineering Effectiveness," *IEEE Transactions on Engineering Management,* vol. EM-33, no. 2, May 1986.
Raelin, Joseph A., *The Clash of Cultures: Managers and Professionals,* Harvard Business School Press, Boston, Mass., 1985.
Shapero, Albert, *Managing Professional People: Understanding Creative Performance,* The Free Press, New York, 1985.
Schneider, Benjamin, "Organizational Behavior," *Annual Review of Psychology,* vol. 36, 1985, pp. 573–611.
Taylor, James C., Paul W. Gustavson, and William S. Carter, "Integrating the Social and Technical Systems in Organizations," in Donald D. Davis, ed., *Managing Technological Innovation,* Jossey Bass, San Francisco, 1986, pp.154–186.
Trist, Eric, *The Evolution of Socio-Technical Systems: A Conceptual Framework and an Action Research Program,* Occasional Paper No. 2. Ontario Quality of Working Life Center, Toronto, 1981.
Westney and Sakakibara, "Designing the Designers," *Technology Review,* April 1986.
Whiston, T. G., *Managerial and Organisational Integration (MOI) Needs Arising out of Technical Change and UK Commercial Structures,* Science Policy Research Unit, University of Sussex, U.K., 1988.

Chapter 3

Methods That Work for Integrating Design and Manufacturing

John E. Ettlie

Introduction

There is substantial agreement among domestic manufacturers today that greater integration between design and manufacturing is a necessary condition to compete in a world economy. In a recent article in the *Wall Street Journal,* Bussey and Sease (1988) wrote that "U.S. manufacturers of all stripes are scrambling to shorten their development cycles and be first to market at home and abroad." Methods called "tiger teams," "parallel engineering," and simultaneous review of designs have cut engineering costs "on an average project 35 percent since 1986" in the case of GM's Chevrolet-Pontiac-Canada Group.

FMC corporation recently reported that "by integrating design, manufacturing and quality...(they) saved more than 40 percent on gauging cost, coordinate measuring machine programming, set-up time and rework" (Burget, 1988). These savings are summarized in Fig. 3.1. What is more, a 50 percent reduction in the direct cost of manufacturing is typical when design is forced to anticipate production (Whitney, 1988).

Most companies want designs that anticipate production and assembly and are now eyeing the entire product life cycle as the locus of integration. Yet there seems to be little agreement on the optimum amount of design-manufacturing coordination, how to resolve trade-offs in design, and how to manage the transition to, and the maintenance of, these new levels of collaboration (Ettlie, 1988).

One sure indicator of the recognition of the problem of coordinating

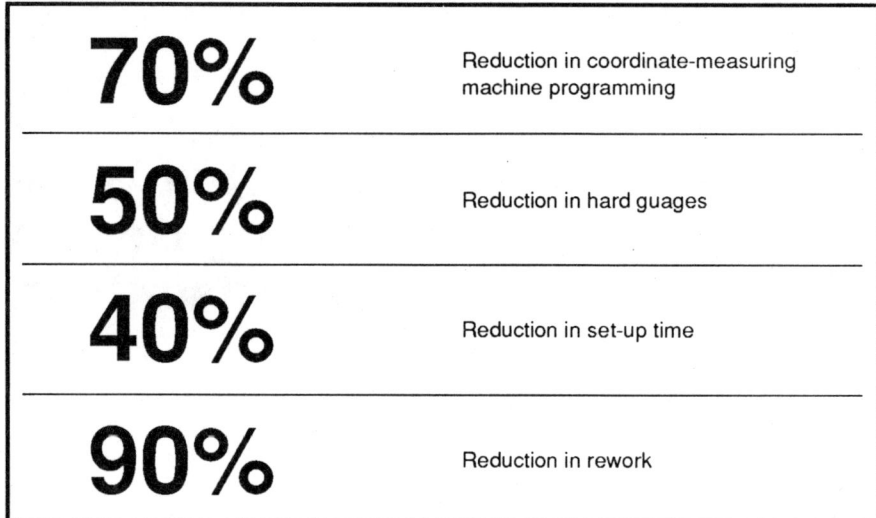

Figure 3.1 Design for quality integration savings at FMC. Eliminating errors of dimensions and tolerances has helped FMC to cut in programming, gauging, setup, and rework. (*Manufacturing Week*, February 8, 1988.)

design and manufacturing is the recent tendency for everyone to say that they have a program in simultaneous engineering. However, when one investigates below the surface of this ocean of verbal commitments, there is often little change in business as usual to substantiate the claims. In any case, this is a sign that firms are now aware that coordination is an important problem. We have found that many people first realized this was an issue when they began to install flexible automation or flexible assembly, found out their parts were not designed to be produced this way, and had to change the parts or the setups on the automation to make accommodations.

In an earlier study (Ettlie, 1986) we found that understanding (or lack thereof) of the dependency between product and process was one of two key factors that managers said influenced the performance outcomes of their modernization efforts [flexible manufacturing systems (FMS), robotics] in the late 1970s and early 1980s. Often products had to be redesigned to accommodate new "flexible" production technologies. The other factor was the quality of the relationship between suppliers of these technologies and plant users.

Kim Clark and his associates at Harvard report (Clark, 1988) that much of the Japanese advantage in bringing products to market sooner lies in organizational structural differences, like the use of "overlapping" engineering and that some U.S. and European firms also use this practice. But one does not change functional organization and tradition over-night. What is more, since many manufacturing firms are also in the process, or at least under pressure to change other things in addition to design or manufacturing, priorities sometimes appear to be shifting rapidly. In order to focus attention on the design issue, experts often argue that cost and characteristics that determine quality in almost an irreversible way are mostly determined once a design is frozen.

Of course, design changes will occur later, but most are minor and the cost of these changes is covered by volume of production after introduction. Anticipating requirements in design and production technology is the biggest challenge yet to be faced by modern manufacturing. In part, the three legs of the stool—Research and Development, Marketing, and Production—are still the paramount concerns in the specification process. That has not changed. What has changed are the methods of integration. Herein lies the first purpose of this chapter.

We draw heavily here from a study to investigate new approaches that domestic firms are using to successfully integrate design and manufacturing during modernization. Herein lies a second purpose, that is really a prerequisite to this task, which is the development of a reliable index to measure design-manufacturing integration. This is a tool that could be used to cut through the fog of words about coordination and be used as a way of starting the change process in a firm and a way of auditing progress along the way.

We base our conclusions in this paper on both the growing literature now beginning to appear on this topic and an empirical study of domestic plants that has just been completed after four years of data collection. We have found that some domestic manufacturers are actually willing to experiment with new approaches like simultaneous engineering to bridge the gap between the cultures of design and manufacturing. They are the few that do more than just talk about it. We also have some preliminary readings on the degree of success of their various efforts. We hope this report will help others put their efforts into perspective.

The Design-Manufacturing Paradox

There are a number of well-accepted reasons why design and manufacturing do not fit like hand and glove without a great deal of effort. The culture of the two functions is different and corporate expectations for the two are not the same. "Doing your job" in design means something entirely different in production. Representatives of the two functions rarely met face to face until recently. But hostile competitive environments have created a new paradox for design and manufacturing. Competition in manufacturing is pressuring firms to shorten their life cycles, which motivates design to try new computer-aided design technologies. They end up owning systems that typically are not compatible with computer-aided design (CAD) systems owned by manufacturing and tooling engineers. Manufacturing and industrial engineering are usually located at plants rather than at headquarters. Automating the design process made things temporarily *worse* in most firms.

But even these trends in CAD adoption are not the real paradox today. The real apparent contradiction is that while design and manufacturing attempt to shorten the product life cycle capability, the goal is really to lengthen product life. Better products have longer design life and better, flexible, integrated design-production systems have longer useful life as well. Many manufacturing firms have forgotten this simple goal.

In a review of the literature and some of the published cases on design-manufacturing engineering including simultaneous engineering, Ettlie and Reifeis (1987) found that teams were the most frequent type of administrative mechanisms used to promote functional coordination of these groups. However, when one correlates the report of team usage with other characteristics of firms, plants, and modernization programs, the result is a blank. That is, team usage does not, in and of itself, distinguish these cases (Ettlie, 1987). This suggests that the characteristics of teams and an inventory of behaviors that results from team usage or other mechanisms for integration would be the more fruitful avenue to pursue. In part, our efforts to develop a second practical measure of *satisfaction* with the design-manufacturing integration process were very revealing vis-a-vis this issue of "teams." We discuss the integrating actions first and take up satisfaction with this process of integration as a separate issue later.

The Five Key Integrating Actions

We visited a total of 39 randomly chosen domestic plants in 1987 that were in the process of modernizing their facilities and interviewed managers, operators, and engineers. We discovered that there were five key actions that summarize the range of practice these domestic plants are using to promote design-manufacturing coordination during the modernization process. These five integrating behaviors amount to a cohesive method (see Appendix 3.1) of implementing a new philosophy of integration of these two organizational functions. The five actions are as follows:

1. All members of the cross-functional team are trained in Design-for-Manufacture methods.

2. Manufacturing signs off on design reviews.
3. Novel organizational structures are used for coordination (detailed later).
4. Job rotation is practiced in engineering functions.
5. Personnel move between engineering and manufacturing—permanently.

A formal method of auditing your operations for this integration approach is presented in Appendix 3.1. The most important and complex action involves the use of new organizational structures to enhance integration, which are presented in greater detail below. We believe job rotation and personnel transfers are actually a transitory stage of action. This activity is a precursor to another alternative, which is the assignment of multiple, permanent jobs to individuals—especially engineers, who are becoming generalists in the organization. Further, we have found that manufacturing engineers are the most likely to be promoted to management during modernization. First we take up the impact of integration on performance outcome and the use of existing organizational structures to enhance design-manufacturing coordination.

Outcomes of Successful Design-Manufacturing Integration

Many managers would take it on faith that at least some of these deliberate actions and policies, like training team members in design-for-manufacture, would result in greater coordination. However, we did not assume it, and what is more, we were interested in the more exacting returns on such activities. Here is what we found. As more of these five integrating actions are adopted by the firm in question, the more likely the following will occur:

- Improvement in cycle time and throughput
- Substantial downstream benefits in reducing the cost of quality
- Significantly higher utilization of the new production system

With respect to the enhanced utilization that results from integrating actions, we found that adopting each additional action (in any order) yields an additional 4.8 percent in linear increase of new system utilization. This result is shown graphically in Fig. 3.2. The score for the action item we determined by assigning a metric of 3 for actual use of the policy or action, assigning 2 to the cases that were in the process of adopting the practice, and using the score of 1 for nonuse

58   Issues and Opportunities

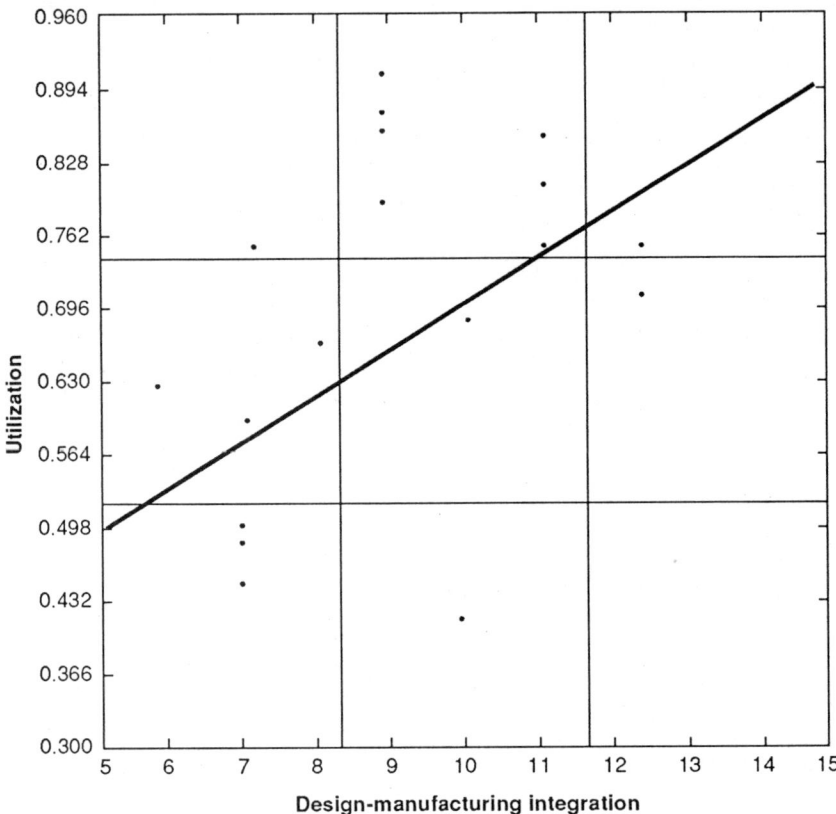

Figure 3.2 Design-manufacturing integration (5 key actions) and utilization of new production systems.

(see Appendix 3.1 for details). The sum of these scores (minimum = 5, maximum = 15) appears on the horizontal axis of Fig. 3.2 and the vertical axis shows system utilization percentages based on a 2-shift basis. There is a nice, linear relationship in these data points. That is, as more of the five key integrating actions are adopted, plants enjoyed higher utilization of these new production systems. In some cases this percentage was in the high 80s or low 90s range.

Ratio of Manufacturing Engineers to Design Engineers

We were very interested in the relationship between the sum of the five key actions for design–manufacturing integration and what plants reported as the ratio of manufacturing engineers to design engineers (average = 1:2.88, n = 27). We found no simple linear relationships between these two measures. However, the results of a test for a curvilinear relationship did turn out to be significant. In this case, the amount of change in design-manufacturing integration actions goes from 3 percent to 49 percent by knowing the manufacturing engineer to design engineer ratio on a simple inverted U-shaped curve. The plot of these two measures is presented in Fig. 3.3.

Based on an examination of Fig. 3.3, it appears that the ratio of design-to-manufacturing engineers that tends to maximize the scores on the design–manufacturing integration scores is between 1.9:1. That is, for this sample of complete data cases (n = 20), the plants scoring highest on novel, cohesive, and successful methods to integrate design and manufacturing had about two design engineers for every manufacturing engineer. This is somewhat below the sample mean ratio of 3:1 design engineers to manufacturing engineers.

It would be an unjustified leap to conclude that this ratio optimized integration. But it does suggest the conditions that lead to innovative practice in domestic plants and firms, beyond the chance level. Note that very low scores on the design-integration scale appear for extremes in the ratio of manufacturing to design engineers, for example, 1:1 and 1:8. The equality ratio extreme may result because so few firms actually have parity in design and manufacturing engineering. But the curvilinear relationship appears to reflect real practices and conditions surrounding modernization.

Structuring for Design-Manufacturing Integration

How does one go about changing organization structure to ensure that good performance tendencies are maintained, and also to encourage higher levels of coordination between design and manufacturing? Is there any evidence that these new mechanisms work?

Common reporting for design and manufacturing engineering

We asked several questions in these modernizing plants about the existing, traditional structure of engineering and production and found some interesting tendencies. These results are included in the summary of Fig. 3.4.

The most typical response (24 cases, 72 percent of the time) represents the firm that does not have design and manufacturing engineering report directly to the same level and has manufacturing engineering report into manufacturing. In most domestic firms, if you were to walk in the door and ask to be directed to engineering, you would not find any manufacturing engineers, only design engineers. Design and engineering are taken as synonymous.

60 Issues and Opportunities

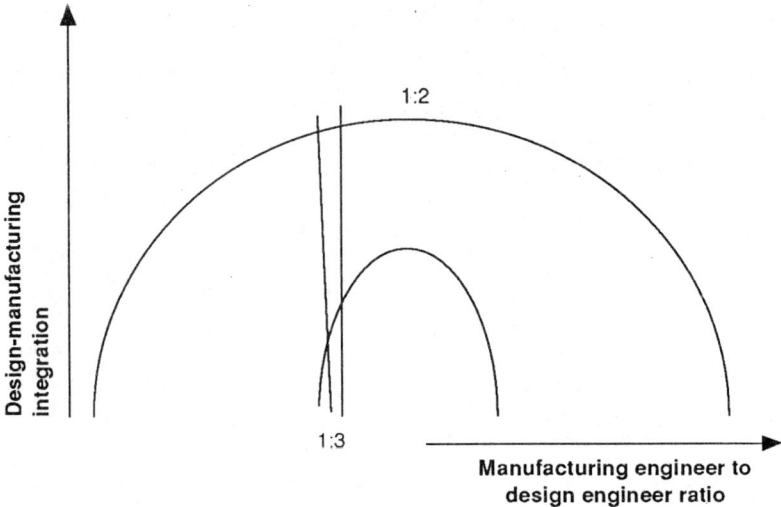

Figure 3.3 Design-manufacturing integration versus manufacturing engineer to design engineer ratio.

The rare case—only four modernizing plants have firms structured this way—is when design and manufacturing engineering have the same direct report, and manufacturing engineering does not report into the manufacturing structure. Of the four cases in our study, three were replacement plants, and very recently modernized. It seems that the departure from past structuring tendencies is just beginning.

We found no measurable advantage to the structural alternative of having design and manufacturing engineering report to the same position for modernization, per se. However, we did find that it depends on how you organize this reporting relationship when you use it. We correlated the various reporting relationships for the 12 cases that

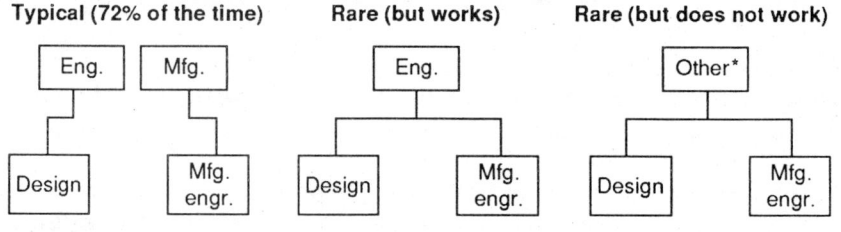

Therefore, recommended: use special structure-not traditional (43% of cases)

---

* Special position, or divisional managment, etc.

Figure 3.4 Common report structure for design-manufacturing integration.

had common reports with performance outcome (see Fig. 3.4). The engineering report is clearly the most desirable option.

Not only did a common engineering report reduce unanticipated expenses, it also produced greater uptime. The number of reporting cases is still low for a common report option, but these trends are interesting and have been verified by a number of anecdotal reports by managers.

Novel organizational structures for design-manufacturing integration

In many cases where firms reported no direct common report for design and manufacturing engineering, they were still experimenting with ad hoc, program-specific organizational forms. It should be recalled that a significant minority (17 or 43 percent) of plants we visited said they had adopted some new organization structure to integrate design and manufacturing for modernization. This action is one of the five key integrating practices we found. We take up the most important nature of these new structures next.

We set out to qualitatively describe what the nature of the new or novel organizational structures was that had been adopted specifically to enhance design-manufacturing integration during modernization programs. Interviews included extensive questioning on the point of the nature of new structures of this type and we summarize the results of what managers told us in these plants in Table 3.1. As we expected, based on the results of earlier investigation on this topic (Ettlie, 1987), the use of cross-functional teams was the most common new structure used. The cross-functional teams used here include the following:

- Design change committee

TABLE 3.1 Structural Adaptation to Manage Simultaneous Engineering

Category	f (%)
Cross-functional planning team	7 (41%)
"Dotted" line report for new position	4 (24%)
Formalizing approval process	2 (12%)
"New" advanced mfg. position	2 (12%)
Business unit on shop floor	2 (12%)
Totals	17 (101%)

- Modernization group
- Interfunctional coordination team
- CAD/CAM steering committee
- A group to consolidate programming
- The adoption of the team philosophy generally and in design

A total of seven (41 percent) of our reporting cases used cross-functional teams to coordinate design and manufacturing during modernization. Next most frequent was the use of "dotted" line relationships on organizational charts—the coordinating relationship formalized—in these domestic plants. Only 4 of the 39 plants actually used this approach, however, which suggests that perhaps much of the coordinating is often done informally rather than formalized on a chart. Dotted line relationships included the following forms:

- Decentralization of manufacturing engineering from corporate group to plant
- New technical jobs on the FMS team
- Involvement of hourly employees in review of product designs
- Research engineer (design) reporting dotted line to plant manager

Formalizing the approval process, using new advanced manufacturing positions—often for engineers, and using shopfloor business units were all reported as being used at least by two plants each.

Payoffs from novel structures for integrating

We examined the relationship between using these new coordinating mechanisms for integrating design and manufacturing during modernization programs in domestic plants and various performance outcomes. We found four significant relationships. The more likely that a plant uses the popular choices for new structures (cross-functional team and dotted line relationship for new positions), the significantly more likely that the following outcome will be reported:

1. Much better coordination of design and manufacturing, overall
2. Greater part family flexibility (more part families scheduled on new system)
3. Faster changeover flexibility (median changeover = less than 1 hour)
4. Greater return on investment (ROI)

This is the first pattern of structural adaptation that we have found that is significantly related to business outcomes for modernization. The relationship between novel structures and ROI is presented in Fig. 3.5 as a cross-tabulation. Although only 14 cases have valid data for both of these measures, novel structure and ROI, the results are still significant. Return on investment ranges from 1 to 119 percent; the worst a cross-functional team did was 15 percent ROI, and the rest of these cases are clustered at the high end of the ROI graph along with dotted line adaptations. With few exceptions, the more innovative the structure, the better the ROI. One can account for about 17 percent of the variance in ROI for these projects by knowing which organizational mechanism was used to coordinate design and manufacturing.

Team composition

We were curious as to the composition of these cross-functional teams. We knew design and manufacturing were represented (the latter usually was the manufacturing engineering department), but we were also wondering what other disciplines and functions were involved in design decisions.

The results of our questions on this topic of team member functional attachment appear in Table 3.2. Of the 25 cases reported, 9 (36 percent) came from marketing. This is a rather significant finding, be-

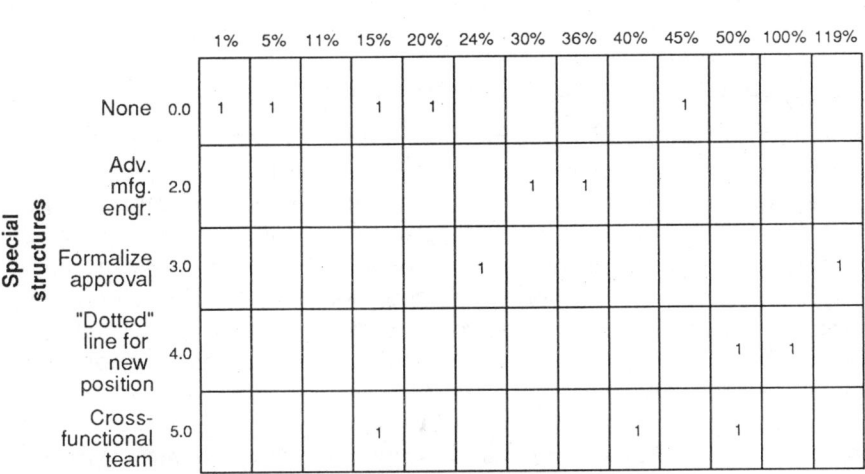

Figure 3.5 Cross-tabulation of special structures to manage simultaneous engineering and ROI of projects.

TABLE 3.2 Cross-functional Team Members Representing Areas Outside Design and Manufacturing

Function represented	f (%)*
Marketing (sales)	9 (36%)
Quality (rel. engr.)	6 (24%)
Business unit or corp.	4 (16%)
Service	2 (8%)
Other [Info. systems, finance, customers, purchasing (one each)]	4 (16%)
	25 (100%)

*Percentages based on total number of functions mentioned rather than number of cases (39) or responses (15).

cause we have found that most companies are struggling with the challenge of how to incorporate the voice of customer into the design process. Here we see one answer to this problem—involve marketing or sales directly in the team process. What is more, quality was the next most frequently mentioned function represented with six (24 percent) of the cases responding so. Perhaps the biggest surprise was the infrequency with which purchasing was said to have participated in the cross-functional team. This raises significant questions about the way domestic plants currently organize for modernization, given the increased tendency to outsource and involve suppliers in decisions. It also represents a significant strategic opportunity, it would appear.

Satisfaction with the Process of Design-Manufacturing Integration in a Unit

We developed a reliable and useful measure of satisfaction with the overall process of integrating design and manufacturing. An audit tool is included in Appendix 3.1 that can be easily self-administered in most firms. This measure has strong overtones of group and team cohesiveness. Satisfaction would not be expected to be directly related to performance outcomes, especially in the short-run. On the other hand, it might be expected to be related to personnel turnover or perhaps absenteeism. A summary of the elements of this satisfaction measure appears in Fig. 3.6.

As one can see, this satisfaction index includes some rather important indications of how things are going during this transition process to simultaneous engineering. It includes the all important notion that *groups* are responsible for coordination, not individual managers or leaders. People are obviously trading off leadership depending upon

Methods for Integrating Design and Manufacturing 65

- Have a D-M team
- A group coordinates D-M
- Team includes members outside D-M
- New policies to integrate D-M
- Line item for manufacturing R&D
- D-M work together during design
- Success of integration effort rated high

Figure 3.6 Satisfaction with the design-manufacturing (DM) integration process.

the issue. The wrenching process of adopting a new philosophy cannot be taken lightly. We know of several examples of people who have lost their jobs in the due course of events, primarily because they simply could not adapt to the group change process. Team membership is also rated high on this index as would be expected. But we have found that every team is different and has its own unique personality. This requires thought and in-depth reasoning to be successful.

Policies for integration

The index includes a cryptic item: "New policies to integrate design-manufacturing integration," which requires some explanation. The actual answers to these questions (after "yes" or "no" are established) have been summarized in Table 3.3.

The new policies that people are talking about here include very specific, general manager directions that are really philosophical shifts in firm orientation. For example, the term "contract specific design," implies that every design decision is made now with the agreement that consensus is necessary for progress. The other example given, "policy of contribution," is also another wonderful general manager directive. Now everyone has to have a role in the design process

TABLE 3.3 New Policies or Practices to Integrate Design and Manufacturing

Category	f (%)*
Policy to integrate functions (for example, contract specific design; "policy of contribution")	10 (53%)
Push CAD/CAM integration**	3 (16%)
Engineering change request (ERC) policy changes	2 (11%)
Formalize design reviews	2 (11%)
Concurrent engineering program	2 (11%)

*Percentage based on mentions, not cases.
**Median percentage of production CAD/CAM integrated = 17% (n = 20).

and no function can be left out. This is not just a matter of making people feel good—it goes well beyond that. People have a contribution to make and designs are diminished if they do not represent the "corporate" thinking that is best on a project.

Satisfaction and performance

We analyzed the relationship between these satisfaction scores for design-manufacturing integration with the other factors of interest. Although we expected satisfaction with the integration process to be significantly related to turnover reduction on the deployment team, there was only a moderate tendency for this to occur in plants. Since this was not an exercise in measuring overall job satisfaction, this result is not really a problem. Many engineers (especially manufacturing engineers) are transferred during projects, as well, so the turnover measure may not be as accurate as one would like.

We did find that design-manufacturing integration process satisfaction was significantly higher in larger plants (500 or more employees), and also directly related to the tendency to share power during modernization. This latter tendency often included such items as using a technology agreement with the union for modernization which broadens job categories.

We also found that satisfaction with the design-manufacturing coordination efforts of a unit was significantly related to one performance outcome: the degree to which modernization improved sales or increased market share. So it is clear that satisfaction with this process is important, as most managers would suspect.

As we anticipated, satisfaction with the overall integration process and the actual level of coordination (five key coordinating actions) were *not* significantly correlated, although the direction of the relationship was positive; which is what other studies have found for the general level of job satisfaction and performance (Majchrzak, 1988). There is obviously more to functions getting along than just coordination. There is also the issue of how people *feel* about this process.

Successful Design-Manufacture Integration

How does one know when progress toward an unprecedented goal is being made? The challenge of integrating design and manufacturing is an excellent example of the more general issue of finding good "in-process" measures of success. What milestones might be good indicators of subsequent success in achieving a sustainable cooperative relationship between these two complex functions—design and manufacturing?

We asked our domestic plant study respondents the following question: "What measures do you use to evaluate the success of design-manufacturing integration"? The results of categorizing the answers to this question are presented in Table 3.4.

The majority of our respondents (20 or 51 percent) said that ultimate outcomes are the proof of the pudding. That is, the only real way to tell if you have had successful design-manufacturing integration is to see if the project was a success. Was it timely, did it meet cost goals, was it helpful in improving overall quality or did the new product-process combination meet quality goals, and so forth. Productivity and reduction of change requests were also mentioned in the ultimate outcome category. We want these types of things to happen, but this type of answer doesn't really help us manage the integration process.

What is perhaps even more discouraging, but real nonetheless, is that the next most frequent response in Table 3.4 is that the "organization does not evaluate design-manufacturing integration" at all. This is probably because these 10 (26 percent) firms in the sample do not really exert an effort that could be considered extraordinary in trying to improve design-manufacturing coordination.

The third category on Table 3.4 does address the in-process measure issue directly, but note that only five (13 percent) of our respondents said they were doing anything about finding ways to benchmark the design-manufacturing integration process. These in-process measures included *part count reduction, standardization of parts and tools* and, interestingly, the *ability to by-pass vendors*. This last indicator is a good one because it illustrates how in-process measures can still be strategic. This measure is tantamount to saying that we are going to change our make-buy relationship with vendors as a result of this process. This is significant and is also illustrated in the Quad-4 high performance engine case from GM presented later (Chap. 6).

We have obviously just begun to scratch the surface on in-process

TABLE 3.4 In-process Indicators of Successful Design-Manufacturing Integration*
What measures do you use to evaluate the success of design-manufacturing integration?

Category	f (%)
Ultimate Outcomes (for example, cost, quality, productivity, change requests, leadtime, and so forth)	20 (51%)
"We don't evaluate D-M Integration"	10 (26%)
In-process Measures (for example, number of parts, common parts, tools, ability to bypass vendor)	5 (13%)
Missing	4 (10%)
Totals	39 (100%)

SOURCE: NSF study data from J. E. Ettlie of 39 domestic plants collected in 1987.

measures of integration. This should be a key issue for researchers and practitioners in the next five years.

Summary

We have discovered that there is an internally consistent and practical way to gauge the degree to which design and manufacturing engineering deliberately attempt to coordinate their efforts for new product and process introduction in domestic manufacturing (Appendix 3.1). Not every firm does everything right, but all of the salient integrating actions contribute to better integration between these two important functions. These actions include training team members in Design for Manufacturing (DFM), having manufacturing sign-off on design reviews, using novel structures for coordination, and rotating people, both temporarily and permanently, between functions.

When these types of administrative changes are made, even when they are difficult and take time to implement, large change programs for processing technology are more successful. In particular, better cycle time and throughput result, significantly higher utilization of new systems is obtained, and downstream cost of quality reductions occur.

The ratio of design to manufacturing engineers in the successful integrators among domestic plants is about 2:1. The average was about 3:1, which doesn't sound like a big difference but it is. Ratios of 8:1 produce the worst integration results, independent of location and cultural differences. We found no evidence that the source of design change requests has any visible bearing on outcomes, either in a linear way or curvilinearly.

With respect to organizational structures for enhancing design manufacturing coordination, most firms use novel, ad hoc structures, rather than relying on existing structures or permanent changes. When design and manufacturing engineering do have a common report (very rare) it is best to have these two groups report to engineering rather than manufacturing or some other function.

The most common special structures for coordination are cross-functional teams and dotted line relationships for new positions in the plant or business unit. When this approach is used, better integration results, the systems implemented are more flexible (part families and changeover time) and there is significantly higher ROI for the modernization project.

When cross-functional teams include members outside engineering and manufacturing, marketing and quality are most likely members. Having outsiders enhances the satisfaction with the overall process of integration.

We have found an internally consistent measure of satisfaction with the design-manufacturing integration process, and it is not necessarily correlated directly with the integration effort itself, that is what you are doing to integrate, or the productivity outcomes of integration—with one exception, increased market share. Satisfaction can stand alone, by itself, as a separate issue.

This measure of satisfaction with the process of integrating design and manufacturing has seven indicators (see Appendix 3.1):

1. A team forms to integrate design and manufacturing.
2. A group is responsible for integration, rather than a single individual.
3. The team includes representatives from functions outside design and manufacturing, like marketing.
4. New policies are being used to coordinate design and manufacturing.
5. Manufacturing R&D is part of the budget (not an expense).
6. Design and manufacturing engineers work together during the design phase of the project.
7. The overall process of integration has been successful.

This satisfaction metric could be used to predict the long-range stability of change efforts in the organization as a whole, and in larger plants, the propensity that other structural changes will be made to decentralize the change process. The plants that were more satisfied with the process saw better market impact of their efforts from modernization.

Cautions for Reorganizing

There are several issues that come up during modernization that highlight the problems that design and manufacturing have in coordinating their efforts. When stress is high and a lot is at stake, mistakes are possible and costly. Some of these mistakes are the result of lack of foresight in the crucial area of predicting the outcome of a new project before it gets started. We have given some action guidelines here based on the experience of others but a few cautions are in order.

First, we have not documented a single case of engineering chargebacks in any of our cases to date. This practice is just now beginning to be used, and we suggest caution. When engineers and design teams take warranty responsibility for their actions they also need to have *all* the information and authority to influence design at the outset of this practice and general accountability of engineers is going to be used.

Second, the choice of leadership on teams is still a crucial decision and one where general managers will have their greatest impact on the process. If the team leader(s) are not both technically and managerially competent, the project takes on a new dimension of risk. One thing that could occur is premature commitment to a technology for the project.

Third, and finally, giving any one person too much responsibility in this process undermines both the potential for coordination and "gut feeling" that people have about what makes a good team and a bad team. As long as everyone does his or her job and can get credit for it, an appropriate amount of visibility has been assigned to the program. But recall that the actors are changing in this process. Where an engineering representative sat before, a vendor, skilled tradesperson or quality manager may now be involved. It requires tolerance, vision and patience with this process to make it work. The most innovative firms involve blue-collar representation early in the decision-making process *with* engineers.

Appendix 3.1

The following presents the questionnaire that was used.

Questionnaire: Actions to Integrate Design and Manufacturing

Below is a series of questions and statements that may or may not reflect the process by which your business unit or location attempts to coordinate design and manufacturing, especially when a new product, or new process, or both are being introduced.

All members of your involved groups should be polled separately and then merged. (Cross differences should be noted.)

Answer the questions or indicate the degree to which you use these practices by circling 1 for "yes," 2 for "no," and 3 for "in process." Some questions also require explanation.

	Yes	No	In Process
1. We used a computer-assisted project management program to deploy this system.	1	2	3
2. We have achieved integration of CAD/CAM.	1	2	3
(IF YES) For what percent of production?_____%			
3. We have people who are trained in DFA or DFM.	1	2	3
(IF YES) Who was trained for what?			

_____

_____

_____

_____

4. A manufacturing representative is required to sign off on design reviews for new products on this system.   1   2   3

   (IF YES) When are design reviews signed off?

      a. Concept design

      b. Preliminary design

      c. Prototype

      d. Production release

5. Design engineers work directly with the operating personnel to solve product problems.   1   2   3

6. We have developed and implemented new structures in order to coordinate design and manufacturing.   1   2   3

   (IF YES) What are these structures?

_____

7. R&D for manufacturing is contracted out to other firms.   1   2   3

8. Job rotation between design and manufacturing engineering is practiced in this firm.   1   2   3

9. Manufacturing engineers can make changes in product design.   1   2   3

10. Personnel from design engineering are sometimes moved to manufacturing engineering or vice-versa.   1   2   3

11. Our product and tooling designs are developed simultaneously.   1   2   3

12. We use compatible CAD systems for product and fixture design.   1   2   3

Now recheck all your answers to see that you truly have represented what your organization does in fact do, not what you wish.

See the following discussion for scoring and interpretation.

Scoring the "actions to integrate" questionnaire

First: Merge the answers to the questions from all participating group members.

1. If there is wide discrepancy in answers, discuss differences and note any that are significant for later use. Don't slow down the evaluation process, but don't ignore insignificant divergence either.

2. Agree, tentatively, on a score for each question.

Second:

1. Focus on Questions 3, 4, 6, 8, and 10. *RESCORE* these questions by assigning a value of 3 to "yes" and 2 to "in process" and 1 to "no."

2. Total the scores for these five questions only. This score can range from a 5 (five questions with five no answers totals 5) to 15 (five questions all answered yes, each scored 3, totals 15). REMEMBER, SCORE ONLY QUESTIONS 3, 4, 6, 8, and 10 for this part of the exercise.

Third:

1. Use your total score to compare with the interpretation categories below.
2. Take note of any qualitative information supplied in Question 3 for improvement later—for example, who was excluded from outside training and development.
3. If your group answered "yes" to Question 5, your unit is among the most innovative in North America.

Score	Interpretation
14–15	Your organization or unit is truly unique. You are at the head of the pack in North American Manufacturing in integrating Design and Manufacturing. Your next challenge will be to diffuse this know-how to the rest of your firm and plan for the "design office of the future." (See Chap. 12.)
11–13	Your unit has come far in integrating design and manufacturing. Your next steps are to be sure that other functions like marketing are phased into the process appropriately and that you move to the next level of integration represented by the top group.
9,10	Your unit or firm is average among North American manufacturing based on a sample of firms modernizing their facilities in 1987 to 1988.
7,8	Your unit is behind the pack but has attempted some actions for integration. You have some selling to do to other units and well-placed managers.
5–6	You have your work cut out for you. If at least one top manager or general manager in the unit has not voiced support for the idea of design-manufacturing, *OR* if there is no significant grass-roots support for the idea, it may be difficult to ever achieve even modest coordination.

Satisfaction with the Design-Manufacturing Integration Process in Your Unit

Below is a series of questions and statements that may or may not reflect the process by which your business unit or location attempts to coordinate design and manufacturing, especially when a new product, or new process, or both are being introduced.

All members of your involved groups should be polled separately and then answers should be merged. (Cross differences should be noted.)

Answer questions or indicate the degree to which you use these practices by circling 1 for "yes," 2 for "no," and 3 for "in process." Some questions also require explanation.

	Yes	No	In process
1. We have a design-manufacturing team	1	2	3

2. A GROUP is responsible for coordinating design and manufacturing

(IF YES)

  a. What is the group called? _____

  b. Does the group report to engineering, to manufacturing, or to another unit?

  1. Engineering

  2. Manufacturing

  3. Other: _____

	Yes	No	In process
3. We have a team that is responsible for product design and development which includes representatives from areas OUTSIDE of design and manufacturing, for example, marketing	1	2	3

(IF YES) Who is on this team? We do not need names, only titles.

	Yes	No	In process
4. We have implemented new policies or practices in order to integrate design and manufacturing	1	2	3

(IF YES) What are these policies or practices?

	Yes	No	In process
5. Our design engineers met with the system vendor to obtain advice on process design	1	2	3
6. Design engineers are paid on the same scale as manufacturing engineers	1	2	3
7. There is a line in the budget for manufacturinjg R&D or its equivalent (not an expense item)	1	2	3
8. Design and manufacturing engineers work together in the design phase of a product	1	2	3

9 a. What measures do you use to evaluate the success of desing-manufacturing integration?

_____

_____

b. How successful has this integration effort been
1. Not successful at all
2. Not so successful
3. Somewhat successful
4. Successful
5. Very successful

Scoring the "Satisfaction Questionnaire"

First: Look at your answer, or the answers your group has provided, to items 5 and 6.

1. If the answer to item 5 is "yes," (1 circled) and/or if the answer to item 6 is "no" (2 circled), *DO NOT* go on with scoring. Recycle through the other questions and make sure these other answers are truly accurate. Even if only one of your answers change, it could be significant. (If you answered item 5 "yes" and item 6 "no" your unit is among the rare 30 percent and 13 percent, respectively, of manufacturing firms).
2. Once you are satisfied the answers are accurate, continue on.

Second:

1. Recode all your item scores *except* 9b as follows: For each 1 ("yes") you circled, score a 3. For each 2 ("no") you circled, score a 1. For each 3 (in process) you circled, score a 2 for each respective item. For item 9b, score the response as numbered. For example, "not successful at all" is scored 1; "very successful" is scored 5.
2. After you have rescored items 1, 2, 3, 4, 7, 8, and 9b, add up your recoded TOTAL. Use only these seven items, *exlude* items 5 and 6, they are not part of the exercise.

Third: Examine the answers to the open-ended questions for leads on how to improve the process in your organization—no matter how high you score.

Here is what your total score means (seven items):

Score	Interpretation
23	Your business is truly rare in its satisfaction experience with efforts to integrate design and manufacturing. Less than 10 percent of all domestic manufacturing firms are your peers. You have achieved a very high level of satisfaction (not necessarily performance) with this coordination process.
21 or 22	Your unit has achieved very high levels of satisfaction. You have scored above 75 percent of domestic firms.
19 or 20	Your groups are above average in their satisfaction with the design-manufacturing coordination process.
16 or 17	This is the average and the median for domestic manufacturing firms. About 25 percent of all units fall at this mid-level. Therefore, your situation is very typical and representative of all domestic manufacturers.
12, 13, or 14	Your unit is below average in its satisfaction level with the design-manufacturing integration process. If satisfaction is important or if you are not *early* in the process of change, significant action is probably, but not necessarily, recommended.
7 to 11	Your business unit or groups are among the lowest 13 percent of all domestic manufacturing firms in their satisfaction with the design-manufacturing integration process. Direct, immediate action is probably recommended including using professional facilitation to help group meetings become more productive. Consider locating design and manufacturing engineering together, and so forth.

A note on scale development

In order to establish a valid and reliable instrument for evaluating the status and progress of managing the design process, we went to great lengths to plan for both face and internal reliability of the scales used in the questionnaire presented in this appendix. We began nearly four years ago to review specific behaviors (rather than attitudes or feelings) that appeared to be associated with a good fit between functions that are represented at the core of the planning and implementation process for new and improved designs in manufacturing.

We started with an item pool of over 100 of these behaviors (for example, "we worked together at the design versus the prototyping stage," and "we spend __ percent time on concept development") and had colleagues and managers review this list. We eventually reduced the list that was used in the domestic plant study to 37 questions for our interview schedule. Data collection was completed in 1987 at 39 locations that had recently installed flexible manufacturing and flexible assembly systems. From these 37 items we constructed two scales that are used in the materials above.

The scale to measure *design-manufacturing integration* resulted in the inclusion of five items with a Cronbach alpha = 0.73, an average interitem correlation of r = 0.36 (n = 27), and a scale mean of 9.15 (sd. = 2.92). This mean is lower with five items than the hierarchical integration scale with four items, which suggests that administrative innovations to support design-manufacturing integration, in general, are more rare in domestic plants.

The design-manufacturing items were:

1. "We have people who are trained in DFA (Design for Assembly) or DFM (Design-for-Manufacturing). (IF YES) Who was trained for what?"
2. "A manufacturing representative is required to sign off on design reviews for new products on this system."
3. We have developed and implemented new structures in order to coordinate design and manufacturing. (IF YES) What are these structures?"
4. "Job rotation between design and manufacturing engineering is practiced in this firm."
5. Personnel from design engineering are sometimes moved to manufacturing or vice-versa."

The most rare modernization related behaviors, based on interview reports and examples within this scale, are job rotation between design and manufacturing engineering, with only 3 (7.7 percent) of the plants agreeing with that statement, and the use of training in DFA or DFM with only 6 (15.4 percent) of the plants reporting that practice as part of their modernization program. Movement of engineers on a permanent basis between design and manufacturing is apparently a more common practice, although still relatively rare, with 13 (33 percent) of the plant representatives reporting the practice. Typical examples include two cases of senior manufacturing engineers in automotive plants being transferred to product teams.

The scale used to measure *satisfaction with the design process,* includes the points summarized in Fig. 3.6, and amounted to seven items. The scale had a Cronbach alpha of 0.78, which is marginally higher than the first scale, but both have very acceptable levels of internal consistency based on this coefficient. The average interitem correlation for the satisfaction scale was 0.34. The scale mean (scored

3 = yes, 2 = in-process, and 1 = no) was 16.91 (sd. = 4.34). The theoretical mean would be 14 for a seven-item scale.

Based on corrected item-total correlations, the most representative items of the satisfaction scale were "A group coordinates design-manufacturing," with a coefficient of 0.58 corrected for the total, and "Have a design-manufacturing team," with a corrected item-total coefficient of 0.72 (the highest on the scale). But both satisfaction and integrating mechanism scales with their explanations (what is a new structure and a new policy) need to be taken together to get the full depth of the picture that emerges in a firm when changing to a new philosophy of design.

References

Burget, Phillip, "FMC Integrates Design and Production; Cuts Costs 40 percent," *Manufacturing Weekly*, February 8, 1988.
Bussey, John, and Sease, Douglas R., "Manufacturers Strive to Slice Time Needed to Develop Products," *Wall Street Journal*, February 23, 1988, pp. 1, 16.
Clark, Kim B., Chen, W. Bruce, and Fujimoto, Takahiro, "Product Development in the World Auto Industry: Strategy, Organization, and Performance," Harvard Business School, Working Paper, 1988.
Ettlie, J. E., *Taking Charge of Manufacturing*, Jossey-Bass Publishers, Inc., San Francisco, 1988.
Ettlie, J. E., "Integrating Design and Manufacturing During Modernization," *TIMS/ORSA National Meeting*, New Orleans, La., May 4–6, 1987.
Ettlie, J. E., and Reifeis, S. A., "Integrating Design and Manufacturing to Deploy Advanced Manufacturing Technology," *Interfaces*, vol. 17, no. 6, November–December, 1987, pp. 63–74.
Ettlie, J. E., "Implementing Manufacturing Technologies: Lessons from Experience," in Donald D. Davis, ed., *Managing Technological Innovation*, Jossey-Bass Publishers, Inc., San Francisco, 1986, pp. 72–104.
Majchrzak, Ann, *The Human Side of Factory Automation*, Jossey-Bass Publishers, Inc., San Francisco, 1988.
Whitney, Daniel E., "Manufacturing by Design," *Harvard Business Review*, July–August 1988, pp. 83–91.

Chapter

4

Design for Life-Cycle Manufacturing

Henry W. Stoll

Introduction

This chapter is about a philosophy of product design called Design for Life-Cycle Manufacturing (Design for Manufacture or DFM for short) and strategies and approaches which can be used to help implement it in design and manufacturing organizations. The DFM philosophy may be defined very broadly as "the full range of policies, techniques, practices, and attitudes that cause a product to be designed for the optimum manufacturing cost, the optimum achievement of manufactured quality, and the optimum achievement of life-cycle support (serviceability, reliability, and maintainability). The concepts of Design for Assembly, concurrent or simultaneous engineering (the design of the manufacturing process and the product at the same time), and other systematic design approaches which cause the engineer, from the outset, to consider all elements of the product life-cycle from conception through disposal are specifically included within this definition."*

In this chapter, we first review the concepts upon which DFM is based. This is followed by a discussion of design process improvement from a design perspective (see Chap. 7 for an organizational perspec-

---

*A version of this definition was originally suggested to the author by Bud Rose of Brunswick Corporation.

tive). A variety of DFM approaches are then presented which build upon the earlier material.

Underlying Concepts

Narrowing choices

Design solutions are arrived at by making choices between a variety of possible alternatives. Problems of design, therefore, seldom have one unique answer. Often, the design solution that is eventually evolved depends on the way the problem is defined or specified. The evolutionary nature of design or the art of design is also very influential. Once a manufacturing approach or particular product technology is found that works, no one wants to tamper with it.

Perhaps the strongest determinant affecting the choices that are made is the culture, life experience, world view, and product knowledge brought to the problem by the engineers who do the design. Given the same problem definition, different designers, design teams, or firms are likely to arrive at relatively different solutions depending upon how each chooses to interpret the problem definition and on the methodologies, procedures, and processes of design that are followed.

These considerations lead to the following observations:

1. In general, many different solutions to a design problem are possible.
2. The solution that is chosen may be selected for one or a combination of many right and wrong reasons
3. Unlike the scientific method, the design process provides no intrinsic guarantee that the selected solution is, in fact, the best solution or even the right solution.

These observations raise a fundamental question of design: if many different design solutions are possible, what is the one best solution and how can it be reliably identified in the most reasonable period of time?

Design for Manufacture attacks this question head-on. By superimposing manufacturing and other life-cycle process requirements on the functional requirements, the feasible design region is reduced to include only a small portion of the initial choices (see Fig. 4.1). Because the narrowed region of design choices is relatively small, the best design can generally be found quickly. And, because all needs are included, the best functional design will usually also turn out to be the design that is easiest to manufacture.

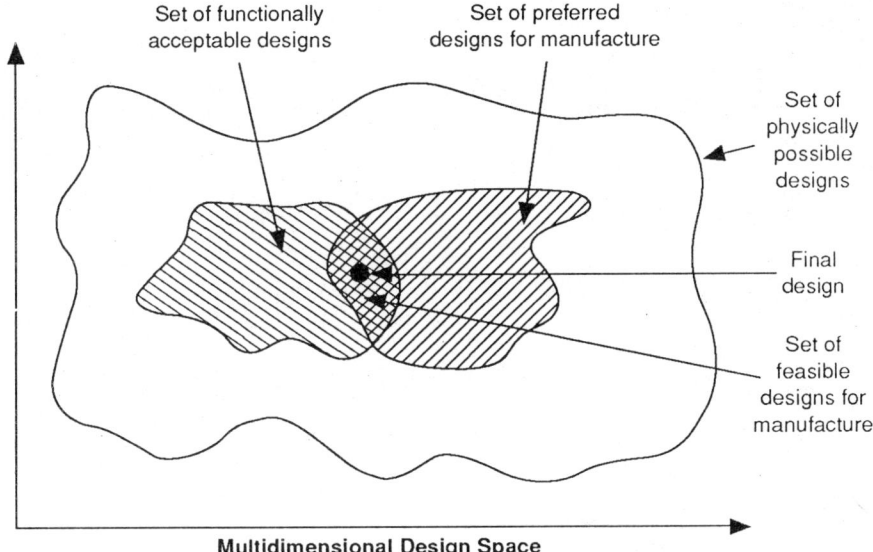

Figure 4.1 DFM rapidly narrows the field of choices to the best few.

Basic precepts

Design represents a progression over time from the abstract to the concrete. The activities involved in this progression can often be divided into a time sequence of phases (Fig. 4.2). As part of each phase of the design process, many questions of design must be resolved and technical and economic decisions made. These decisions generally require a great deal of information and the quality of such decisions often depends directly upon the completeness, correctness, and availability of needed information. If the required information is not available, the designer* makes the best decision he or she can and then reexamines the decision at a later date as more complete information becomes available. This process of reexamination is the iterative nature of design.

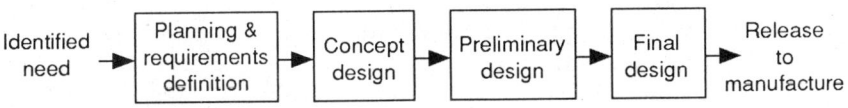

Figure 4.2 Typical phases of design.

---

*Throughout this chapter, the terms "designer" and "design team" are used interchangeably. "Designer" is typically used when discussing the way an individual concerned with design might think or act. "Design team" is used when group behaviors are more relevant.

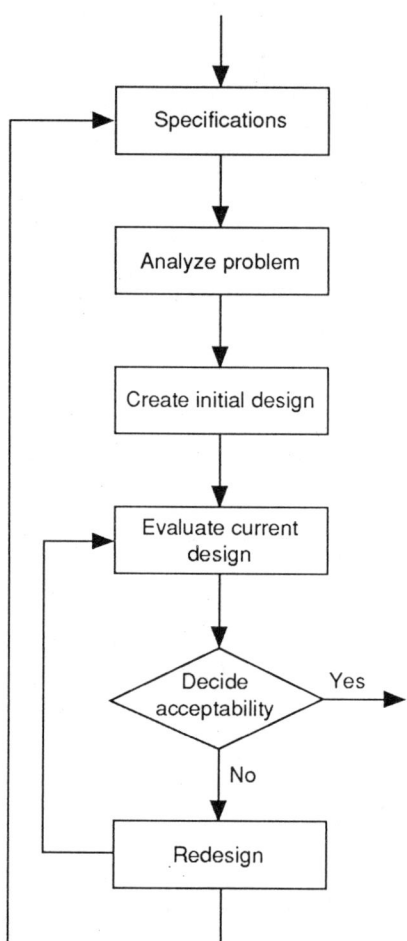

Figure 4.3 Iterative model of the design process. [*Reprinted, from Computers in Engineering, vol. 1, 1984, ASME (Dixon and Simmons, 1984).*]

Design iteration can be modeled as an iterative activity of generating and evaluating candidate designs (Fig. 4.3). The process begins with a design specification being given to the designer. Using both general information and specific information about the design problem, together with product knowledge gained through past experience, the designer first analyzes the problem to find the best way to approach the design and then, based on his or her analysis and the insights gained, generates an initial design. The designer then evaluates this candidate design using the best available engineering practices and methods. Based on this evaluation, the designer makes a judgment regarding the acceptability of the design. If the candidate design is unacceptable in one or more ways, the designer alters the design in an attempt to correct the identified failings. This is the redesign step. The new candidate design is then evaluated and the process repeated until either an acceptable design is found, or the designer concludes that the specification cannot be met as formulated.

Combining the ideas underlying Figs. 4.2 and 4.3, we see that, as choices are made, uncertainty about the design is reduced. At the beginning of a project, the design team must deal with considerable ambiguity. Of the many possible choices, which subset of choices will best satisfy the design objectives within the budget of time and money available? As time into the design increases, the uncertainty is reduced through a series of iterative redesign cycles. Eventually, a particular solution principle and design configuration are selected and attention shifts to the arrangement and detail design of individual parts. The detailed design of each part again follows the same pattern of uncertainty reduction but on a smaller scale. Many material, form, and manufacturing process possibilities must be narrowed down to one particular material, geometry, and method of manufacture jointly optimized to meet the functionality, cost, and production rates specified.

Uncertainty reduction can be visualized as a "funnel process" in which the ambiguity is narrowed over time as questions of design are answered and choices are made (see Fig. 4.4). Iterative redesign cycles (depicted as "curly-ques" in the figure) are the primary mechanism of uncertainty reduction. The funnel process is characterized by two distinct behaviors—flexibility and optimization. In the early stages of design, problem-solving activities tend to revolve around the investigation of alternatives. Many solutions or design directions are possible. The design is fluid and in a constant state of flux. Dealing with this type of ambiguity requires that the design team be flexible in its approach and in its ability to handle change.

As the design ambiguity is reduced, attention narrows to one solution. At this point, the primary problem-solving activities shift from consideration of alternatives to considerations of optimization. Optimization generally involves refinement of a particular design concept or configuration. Also, as shown in Fig. 4.4, optimization is generally carried out within a restricted range of choices.

Viewing uncertainty reduction in design as a "funnel" process helps bring the fundamental nature of the design process into clearer focus. In particular, we make the following observations:

1. In order for the uncertainty reduction process to properly converge to the overall "best solution," the process must begin with as complete and comprehensive product knowledge as possible, formu-

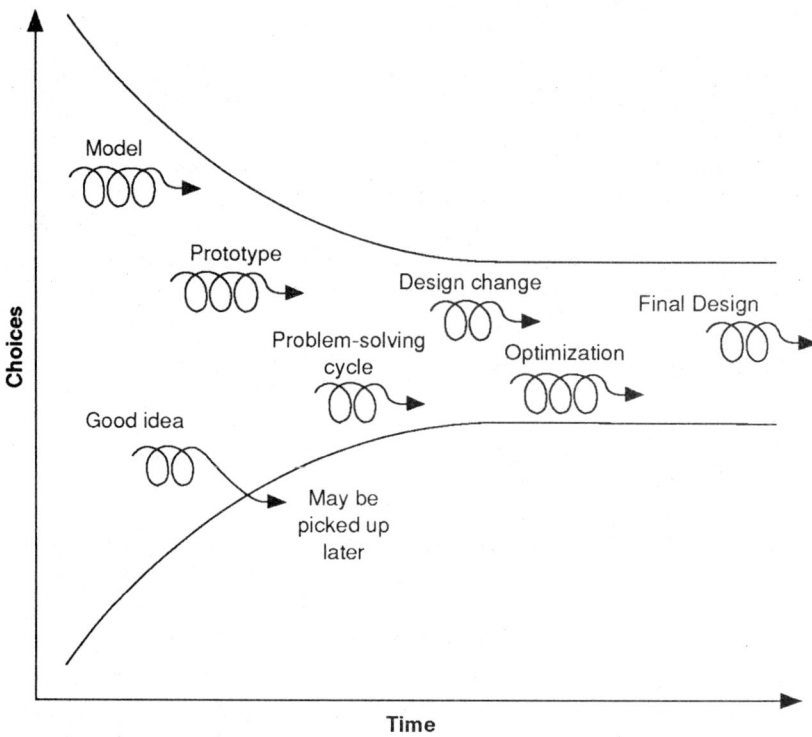

Figure 4.4 Uncertainty reduction modeled as a "funnel process."

lated and clearly stated as requirements and constraints. This means that the "planning and requirements definition" phase (see Fig. 4.2) is *crucial* to successful design.

2. Because the design is fluid and in a constant state of flux during the "funnel" portion of the uncertainty reduction process, the greatest opportunities exist at this stage to "get the design right the first time" and to make choices on a global, life-cycle basis. This is because the very best time to trade-off life-cycle requirements is when conceptual maneuverability is wide. Generally, when the design concept is fairly abstract and hardware is still relatively remote, satisfactory solutions to even serious conflicts between competing requirements can be found. Once the design has been narrowed to one particular solution, the available range of choices often becomes too restrictive to permit effective trade-off and compromise.

3. Design freedom diminishes as the range of choices is narrowed (see Fig. 4.4). Therefore, engineering change, which is inevitable because of the iterative nature of design, is easiest to perform during the funnel portion of the uncertainty reduction process, when many

choices still exist and a variety of alternatives can still be considered. The time to make changes is during the planning stage, not after the detail work has been completed.

These observations form the basic precepts of the DFM philosophy. In essence, DFM is a simple recognition on the part of engineers and managers that each design decision, large and small, carries with it a set of implied manufacturing, producibility, and life-cycle support consequences. Early consideration of life-cycle requirements ensures that the problem of design is correctly and completely defined, that life-cycle requirements are properly considered when conceptual maneuverability is wide, and that the product design is optimized with respect to these processes as well as with respect to functionality and marketability.

Design integration

A product is described in different ways by those who market, design, and manufacture it (Rinderle, 1986). Marketing generally develops a *functional* description that describes the product in terms of target customer needs, problems, expectations, and desires. Engineering develops a second description, consisting of a set of engineering drawings, bill of materials, and other relevant engineering data, that specifies the *form* of the product in terms of geometric information, tolerance, and material types. Typically, this information is supported by engineering prototypes, engineering analysis, and CAD data files. A third description of the product, generated by the manufacturing organization, defines routing sheets, process flow and plant layout, tool and equipment design, work design, part programming, and so forth, required for product *fabrication.* This third description is the manufacturing plan. It is the product description that is used by the operations side of the firm to actually manufacture the product.

These three product descriptions can be viewed as forming three fundamental stages of product development: concept generation, product engineering, and process engineering. A design is complete when all three descriptions have been specified. The means by which the three descriptions are specified is generally determined by the product development process employed. In the United States and many other countries, a serial approach has been followed (see Fig. 4.5). In this approach, each stage is a discrete activity, with each upstream stage an unalterable constraint on downstream stages.

It can be argued that the serial approach has its merits. It helps executives control a potentially messy process, it allows functional managers to focus closely on their jobs, and it allows engineers to be more

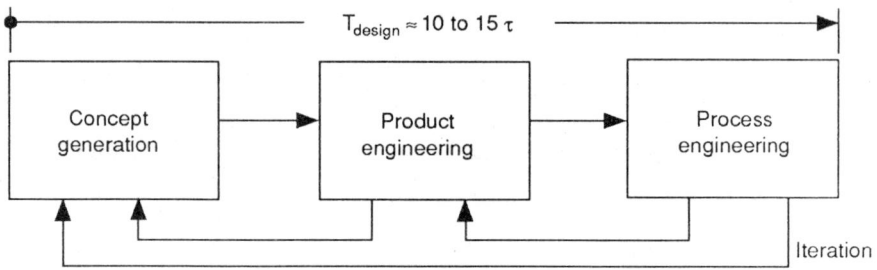

Figure 4.5 Serial approach.

proficient by specializing in particular aspects of the product or by being solely concerned with manufacturing. But, it does this at the price of preventing early consideration of life-cycle process requirements. If we imagine that the development of each product description—function, form, and fabrication—involves a funnel process similar to that shown in Fig. 4.4 placed in series one after the other, we see that both the design choices and conceptual maneuvering room available for follow-on stages can become severely limited by the narrowing of choices performed in preceding stages.

A good example of the problems this can cause is the apparent failure, in some applications, of advanced manufacturing technology to achieve promised productivity improvements. When these failures are carefully analyzed, we often find that the problem is not the fault of the new technology, but rather it is due to a mismatch between the product design and the advanced manufacturing technology employed. Robotics, machine vision, and flexible automation work best when the product has been specifically designed with the needs of these technologies in mind.

Perhaps the most serous drawback of the serial approach is that it often leaves life-cycle cost, quality, and development lead time to chance. By the time problems in these areas are recognized, iteration to fix them is often expensive and time consuming. The result is numerous redesigns, suboptimal and costly total designs, poor response to market and technological change, and excessively long design cycles.

An alternative product development approach is to integrate the development of the three product descriptions into a single engineering effort. Instead of developing each product description in a discrete and separate activity, all three representations are developed concurrently in an iterative fashion that allows for continuous exchange of ideas and timely feedback of implications (see Fig. 4.6). The idea of design integration has become known by several different names. In some circles, it is referred to as the "overlapping approach" because of the way the design stages overlap or proceed in a more concurrent manner

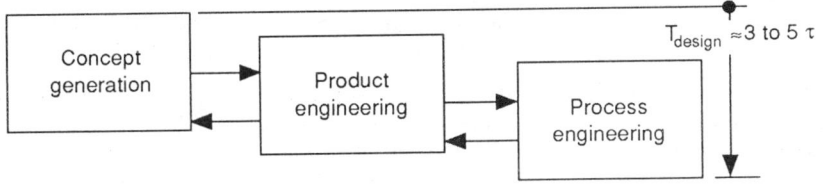

Figure 4.6 Overlapping approach.

(see Fig. 4.6). Other terms which are commonly used include simultaneous engineering, concurrent engineering, and integrated product and process design.

The advantages of design integration are illustrated by the following integrated product and process design project scenario. In this project, a coordinated product and process concept for a mass-produced cooling product was proposed during the conceptual design phase. The product concept consisted of three sub-systems, a freon system, air handling system, and control module, each installed in a molded plastic lower chassis and held in place by an upper chassis. In the associated manufacturing concept, each subsystem was to be built and tested on its own individual assembly loop and then inserted robotically into the lower chassis on the final assembly line.

In optimizing the proposed product and process concept, the design team realized that robotic insertion of the freon system into the base would be a difficult operation to implement because the compressor is relatively heavy compared to the heat exchangers and the system is nonrigid because of the inter-connecting copper tubing that essentially holds it together. The freon system was originally proposed as a separate sub-assembly because the design team did not feel it would be wise to braze the tubing near the plastic base.

Once the problem associated with robotic insertion of the freon system was identified, the design team rethought this idea. The result was an alternative concept in which the freon system is assembled in the base chassis from the start. This solution avoids the insertion problem altogether, as well as the costs which would have accompanied it had the problem been discovered late in the project. As it turned out, only minor changes to the base component and copper tubing connections were necessary to prevent heat damage during brazing. These changes were readily implemented since no design details had been finalized, no components or tools had been prototyped, no expensive finite element analysis had been performed, and no expensive prototype testing or code compliance certification had been conducted.

In further designing the cooling device and its method of manufacture together as a coordinated system, additional opportunities for product and process simplifications were discovered. For example, several fasteners and seals were eliminated and guiding and self-fixturing features were added to various components and subsystems to make automated handling and insertion easier. Design modifications were made to the molded plastic lower and upper chassis which eliminated all camming of molds. These changes not only simplified manufacture, but also resulted in improved air flow through the unit making it possible to down-size the fan motor. However, the smaller motor required a different fan system design and so a second design iteration was initiated.

With each iteration, additional improvements in both product and process were made resulting in a globally optimized design. Because these iterations were performed early in the design, design costs and time requirements were minimal. Once the coordinated product and process concept was clearly defined and agreed upon, product engineering was able to proceed with detail design and manufacturing engineering was able to begin planning the production system and specifying long lead equipment. When problems arose, they were resolved quickly because all parties had a clear understanding of the product and process concept and the trade-offs involved. The transition from development into production flowed smoothly and, unlike past programs, there were no surprises as the project neared completion.

When implemented successfully as in this hypothetical example, design integration offers great potential for dramatic improvements in cost, quality, and delivery because it avoids many of the problems associated with the serial approach. By integrating the development of function, form, and fabrication into a single funnel process, early narrowing of downstream choices and major iterations between stages are avoided. Moreover, time, effort, and money required to solve problems late in the project are saved and suboptimal designs and costly "fixes," often produced by engineering change performed late in the project or after design release, are avoided. Most importantly, an integrated development process makes it easier to adjust the course of the design to accommodate minor design cycle and market perturbations, provides early warning of incipient producibility and manufacturing problems, and facilitates close coordination between all three product representations.

Team approach

Implicit in the concept of design integration is the idea of a "team approach." The product designer who takes manufacturability into account when creating designs will no doubt generate a more cost effective and producible product. But, experience has shown that having a thorough understanding of manufacturing operations and requirements is not enough to fully realize the potential for improved cost, quality, and delivery offered by design integration. To fully optimize function, form, and fabrication, all stakeholders in the life-cycle manufacture of the product need to be involved with the design from the very beginning using a team approach.

In the team approach, interactions between function, form, and fabrication are communicated, coordinated, and controlled in an on-going continuous manner. Function, form, and fabrication solution proposals are openly shared in order to ensure coordination and to avoid conflicts. When conflicts arise, they are resolved in a cooperative manner.

The key to success is an on-going design review dialog between the product designer and the manufacturing organizations that must "live" with the product once it is released by engineering and transitioned into production. Every phase of product development requires an open-minded review of the redesign alternatives proposed. Assurance that all technological innovations, manufacturing variables, and other downstream requirements have been properly considered can only be accomplished by this on-going review or team approach process (see Chap. 3).

Improving the Design Process

Many companies are finding that new design procedures and strategies utilizing disciplined and systematic methods and procedures are needed to effectively implement design integration and the team approach and to streamline the design process. Improvements in the way a company does design can take many forms, but most will usually involve some or all of the following elements:

1. *Structural change:* Organizational and procedural changes are made to enhance communication between functions, to facilitate integration of function, form, and fabrication, and to simplify and optimize work flow.
2. *Compatible CAD/CAM/CAE systems:* Product and process designers work on compatible systems to allow CAD/CAM/CAE information integration.
3. *Focus on change:* Iteration is managed in a fashion allowing for continuous exchange of ideas and timely feedback of implications from the beginning.
4. *Life-cycle design:* Neither product nor process is optimized, rather the manufacturing system is optimized.

Design process improvement focuses on developing and implementing effective procedures and methodologies for design. Achieving a disciplined design procedure can be largely a matter of taking the time to clearly understand the activities involved in performing design and rationalizing these activities into a step-by-step procedure which makes sense in light of the DFM philosophy. Activities that are not needed or have marginal value are eliminated; those that are deemed important are simplified and streamlined. Where possible, the time required to perform key activities is shortened and bottlenecks are eliminated (see Chap. 7).

Ideally, an activity-based design procedure should be simple enough to become habitual in its use and flexible enough to fit the range of design activities anticipated. It should also emphasize integration of function, form, and fabrication in every step. Once such a procedure is available and fully accepted by the design organization, an effective, overlapping design approach will likely develop naturally.

In seeking specific ways to improve the design process, it is useful to note that iterative problem solving in design is similar to successive approximation methods used to solve equations in mathematics. In finding the root of an equation in this way, one first guesses a value for the root and then checks to see how closely it satisfies the equation. A new estimate of the root is then made based on information gained from the previous estimate. The successive approximation process typically continues until a sufficiently close approximation to the root is obtained. Over the years mathematicians have learned much about this process. For example, the process can be made to converge more rapidly by using special methods, such as the Newton-Raphson method, to help select the next approximation or estimate. When more than one root exists, the initial value tried often determines the root to which the process converges. Hence, considerable time and effort can be saved by studying the equation carefully to select the best starting value. Mathematicians also know that if the initial guess is not carefully selected, the successive approximation process may degenerate into chaotic behavior where convergence becomes impossible.

In looking at Fig. 4.3, we see an exact parallel to the successive approximation process. The design specification represents the equation and the design solution represents the desired root. The designer studies the specification and then creates an initial design based on his or her understanding of the problem. The designer then improves the initial design iteratively until an acceptable solution is obtained. Like the successive approximation process, the initial design determines what the final design is likely to be. If the problem of design is not well understood, then a wrong or inappropriate solution may well be developed, that is, the solution converges to the wrong root. Also, the use of special methods such as Design for Assembly, CAD/CAM/CAE tools, and work flow based organizational structures (Chap. 7) help accelerate the rate of convergence to the final solution. Finally, like a poorly chosen starting value for successive approximation, if the initial design concept is flawed, it is possible that the process will never converge to a finally acceptable design. What this means in practical terms is that a marginal or inferior solution will eventually be accepted when the budget of time and money becomes exhausted.

Although our successive approximation analogy is admittedly an over simplification of a very complex process, it effectively points out those aspects of the design process which need to be emphasized if the process is to be improved with respect to design for manufacture. In particular, the following three design process improvement opportunities are available.

1. *Clarify and plan for "downstream" requirements:* Design requirements imposed by downstream processes such as manufacture are often overlooked in the early stages of design. Yet, it is precisely these requirements that determine the practicality of a particular solution. Preparing thoroughly and carefully for design includes understanding and planning for downstream requirements as well as functional requirements. A final design solution is only as good as the requirements definition used in its development. If the requirements definition is incomplete, then a particular design solution might be judged acceptable when in fact it is unacceptable. A disciplined, step-by-step procedure for defining and effectively considering downstream requirements should therefore be an integral part of an improved design process.

2. *Focus on the initial design:* Recognize the importance of the initial design in determining the final design outcome. If the wrong starting point is selected, then either a costly and suboptimal product is eventually produced or the program fails. In general, it is very difficult to recover from a flawed starting point without literally beginning all over again and even then, there is no guarantee that a second try will be any better. And yet, very little time and money is usually budgeted for this most important first step. The importance of the early stages of design should be clearly recognized and continually emphasized at all levels of the organization. Most importantly, sufficient time and resources needs to be budgeted for these up-front activ-

ities to ensure that the one "best total design concept" is identified and selected.

3. *Use design methodologies and tools effectively:* Using design methodologies and tools wisely is the essential third leg of an effective design process improvement strategy. A clear understanding of how a particular methodology or tool is going to be used should be implicit in the design process. For example, if Design for Assembly is to be used, when and how it is to be used and how its results are to be used should be clear to everyone involved. Many design tools and methodologies can be very effective if the correct *management expectation* is effectively communicated and if their use is correctly integrated into the design process.

To illustrate the pitfalls that abound, consider the use of CAD in the design process. As illustrated in Chap. 1, many design veterans argue that, in the "good old days," product and process integration occurred naturally as the *manually generated* design layout was viewed and discussed by members from the various functional areas. In the CAD environment of today, the design layout is often hidden or compartmentalized within the CAD system and cannot be readily viewed and easily commented upon. Hence, the opportunity for product and process integration that occurs naturally when drawings are generated manually is inadvertently lost when the drawing board is replaced by the CAD environment. Similarly, in some cases designers are now rewarded and promoted based on their CAD prowess rather than on their technical ability as designers. These are examples of how a design tool, if not properly integrated into the design process, can impair rather than facilitate successful design.

Suitable step-by-step design procedures can take a variety of forms depending on the nature of the product, the culture of the design organization, the organizational structure, the availability of design tools such as CAD/CAM/CAE, and the particular needs of a given design project. For many design organizations, the best approach to design process improvement is internal development of a design procedure uniquely tuned to the particular needs of the organization involved. Examples of this approach are presented in Part II of this volume. Alternatively, general guidelines such as those published by the German VDI Society for Product Development, Design, and Marketing (Guideline, 1987) can be utilized.

Disciplined design approaches do not need to be overly complicated or restrictive. This is illustrated by a simple design meeting approach being used by Black & Decker Corporation (Bradyhouse, 1988). The focus of this particular approach is to ensure that design for assembly

Meeting No. 1	1. Review product background. 2. Marketing objectives. 3. Expected annual sales. 4. Engineering concept. 5. Manufacturing location and target costs. 6. Review basics of DFA.
Meeting No. 2	1. Review bill of material. 2. Disassemble prototype or determine assembly sequence using exploded view.
Meeting No. 3	1. Set objectives (these should be stretch objectives such as no lead wires, no screws, etc.). 2. Determine if product should be assembled manually, robotically, or hard automated. 3. Issue progress report.
Meeting No. 4	1. Review product for robotic assembly or hard automation depending on conclusion from previous meeting. 2. Identify impediments to robotic or hard automation assembly. 3. Make assignments to team members to work out solutions to various impediments.
Meeting No. 5	1. Review impediment solutions. 2. Brainstorm alternative solutions. 3. Alter design to remove impediment. 4. Issue progress report.
Meeting No. 6, 7, etc.	Repeat Meeting No. 5 cycle until all impediments have been removed and objectives have been achieved.
Final Meeting	Write management report listing accomplishments such as: • Design efficiency improvement • Reduction in part count • Increase in number of sculptured parts • Design improvements due to creative value analysis • Layout of recommended assembly approach and preliminary cost analysis.

Figure 4.7 Typical DFM/DFA meeting agendas. (*Adapted from Bradyhouse, 1988.*)

and manufacture are effectively integrated into an already existing design process. As part of the existing design process, new product development teams meet every four to six weeks during the entire development cycle to review schedule. To perform the DFA/DFM work, one additional day is tacked onto the product development meeting (Fig. 4.7). These meetings focus the team on producibility and other downstream issues and ensure on-going close coordination between function, form, and fabrication.

DFM Approaches

Many new and innovative design strategies and approaches become possible as the DFM philosophy becomes an integral part of a company's design practices and culture. In this section, we present a variety of ideas that build on the DFM philosophy. For the creative manager, these ideas can provide a basis for design process optimization. For the design team, they offer a variety of possibilities for identifying innovative new product and manufacturing solutions and for identifying the best design possible in the least time possible.

Axiomatic design

In axiomatic design, good design is achieved by using fundamental principles or axioms of good design to guide and evaluate design decisions. Extensive examination of successful designs has shown that good design embodies two basic concepts (Suh et al., 1977). The first of these is that each functional requirement of a product, device, or system should be satisfied independently by some aspect, feature or component within the design. The second basic concept is that good designs maximize simplicity, that is, they provide the required functions with minimal complexity. These concepts have been formalized as the following design axioms (Yasuhara and Suh, 1980):

Axiom 1: In good design, the independence of functional requirements is maintained.

Axiom 2: Among the designs that satisfy Axiom 1, the best design is the one that has the minimum information content.

All properties and features of a good design for manufacture can be derived from these two axioms. The axioms tell us that information content of the design will be reduced by integrating functional requirements into a single physical part or design solution, but only if the functional requirements of the design are satisfied independently. If functional requirements are coupled or become coupled in the design, we should decouple or separate parts or aspects of the design solution even if this means adding a part or increasing information content in other ways. Maintaining independence of functional requirements allows each design problem to be solved independently of the others. This simplifies the design problems that must be solved, enables the use of "tried and true" solutions in new applications, and is often essential for acceptable product operation and performance. Most importantly, when functional requirements are satisfied independently, "ripple effects" are short circuited. Design changes affecting one functional requirement do not affect other requirements. Similarly, variation in hard-to-control parameters that affect one function do not degrade other functions.

One way to visualize manufacturing information content of a particular design is to mentally create routing sheets for the product's manufacture. That is, imagine the number of separate activities and the number of instructions per activity required to manufacture the particular product or subassembly. The best design would be the alternative requiring the least number of activities with the fewest instructions per activity. For example, a product composed of few parts would require less manufacturing and assembly activities than one composed of many parts. A part designed so that it is processed on only one surface would require fewer instructions to fabricate compared to one having features on several surfaces and requiring several re-orientations during fabrication. A dimension that is consistently on target from part to part would require fewer instructions to deal with than one that varies widely or is consistently off target.

Use of the design axioms in design is a two-step process. The first step is to identify the functional requirements (FRs) and constraints. Each FR should be specified so that the FRs are neither redundant nor inconsistent. It is also useful in this step to order the FRs in a hierarchical structure, starting with the primary FR and proceeding to the FR of least importance. Once the functional requirements and constraints are specified for a given product or design problem, the second step is to proceed with the design, applying the axioms to each design decision.

Eliminate, simplify, standardize where possible

This DFM approach seeks to minimize manufacturing information content of a product design to the fullest extent possible within constraints imposed by functionality and performance. Although an eliminate, simplify, and standardize strategy is applicable everywhere in manufacturing, our interest in this section focuses on the following DFM objectives:

1. Minimize the total number of parts
2. Simplify the design to ensure that the remaining parts are easy to fabricate, assemble, handle, and service
3. Standardize where possible to facilitate desirable producibility characteristics such as interchangeability, interoperability, simplified interfaces, effective consolidation of parts and function, availability of components, and so forth

"The ideal product has a part count of one" (Huthwaite, 1988). A part is a good candidate for elimination if there is no need for relative motion, no need to be separate to facilitate assembly or subsequent adjustment between parts, no need to be separate for service or repair, and no fundamental reason for materials to be different. Perhaps the most effective way to eliminate parts is to identify a design concept or solution principle that requires few parts. Replacing a four-bar linkage (four parts) with a cam and follower mechanism (three parts) is a simple illustration of this approach. Integral design, or the consolidation of two or more parts into one, is also highly effective. Integral design reduces the amount of interfacing information required, and decreases weight and complexity. One-piece structures have no fasteners, no joints, fewer points of stress concentration, and often can be sculpted to better utilize material.

Another viable approach to part count reduction is to design multi-use or "building block" parts that can be used interchangeably in a variety of different products, product models, or applications. For example, with the right standardization scheme, the same mounting plate can be used to mount a variety of different components. Multi-use parts reduce manufacturing information content by reducing the number of different parts or part variations that need to be manufactured. They also produce economies of scale because of increased production volume of fewer parts and economies of scope because the same part is being used in a variety of applications and products.

Simplicity of component and assembly design ensures easy fabrication, assembly, testing, and servicing. The first step in achieving a simple design is to develop a systematized product structure which standardizes the relationship between product function, form, and fabrication. Examples of product structure include building block approaches, modular designs, use of a layered or stacked construction, or the use of a base component which locates, orients, and relates the various components of the product to each other. Once a carefully thought out and planned product structure has been formulated, a simple design configuration can often be achieved by implementing the following guidelines:

1. Minimize assembly directions and reorientations. If possible, develop a top-down manufacturing approach. Top-down, z-axis assembly is especially important. Extra directions mean wasted time and motion, increased complexity, and added quality risk.

2. Minimize part variations such as the number of different types or sizes of screws used. Use off-the-shelf or previously designed components where possible. Avoid specials.

3. Avoid separate fasteners. Separate fasteners increase assembly complexity, add extra parts, and create quality risks.

4. Where possible, make parts easy to assemble by providing easy

access; generous tapers, chamfers, and radii for easy insertion and guiding; and self-fixturing features for easy orientation and error free assembly.

5. Make parts easy to handle and orient by providing symmetry and easily identified features, by avoiding very large or very small sizes, and by avoiding features that cause parts to nest, tangle, or become interlocked.

6. Eliminate or simplify adjustments when possible. Identify critical dimensions which, if not confined to a single part, require slots and other features which permit adjustment between parts. If possible, incorporate such dimensions into a single part.

7. Plan the interfacing and layout of electrical connections, flexible tubing, and control cables early in the design to minimize the number of flexible components, simplify assembly, ease servicing problems, and avoid system integration problems that can otherwise occur late in the project.

8. Avoid randomness in the design by providing cable runs, eliminating dangling connectors and unrestrained parts, and providing self-locating features on mating parts to eliminate ambiguity. Avoid part to part or part to process dependencies such as "fit at assembly."

9. Avoid uncertainties in the design. Avoid design alternatives that require specialized or unusual operator skills or training to manufacture, operate, or service. Use components and devices for which well documented information on failure rates and derating is available. Design to avoid or minimize the number of wear surfaces and rubbing parts. Avoid designs that are sensitive to hard-to-control factors.

10. Develop an easy to service and maintain design. Avoid design alternatives whose repair involves hard-to-control factors. Design so that limited life components such as fuses and filters, as well as fluid couplings, seals, and other parts that are at above-average risk of failure are visible for inspection and are accessible for easy scheduled maintenance, removal and re-installation. Provide sufficient hand and tool manipulation clearance for easy maintenance, adjustment, and measurement without removal of interfering components. Ensure that all major components are easily identifiable by serial number or part number without removal or reorientation.

In implementing the eliminate, simplify, and standardize approach, it often proves useful to set "stretch" design objectives early in the design. Examples might include "no separate fasteners in final assembly," "no reorientations of the build during final assembly," or "no tension springs." One way to develop a realistic set of stretch goals is to carefully list the advantages and disadvantages of product design and manufacturing methods used in an existing or similar current production product. Input from factory and assembly workers can also be very valuable in doing this because these people are most knowledgeable about existing design-related production problems and can often offer excellent advice about how to correct or avoid a problem.

Activity–transaction-based accounting systems, when they are available, often provide tremendous insight in helping design engineers identify ways to reduce product cost and set stretch objectives for new designs (Huthwaite, 1989). Unlike traditional labor-based cost systems, activity-based systems allocate overhead costs on measures of activity. For example, purchasing costs are allocated to products on purchase orders, setup costs are allocated on production changeovers, and receiving costs are allocated on receipts. In using such cost systems, designers typically discover that they pay an enormous cost penalty for incorporating any unique low volume parts in their designs. Hence, a stretch design objective arising out of this realization might be to avoid unique parts or parts requiring special processing.

Another way to set stretch objectives is to reverse-engineer leading competitor products. If a particularly clever or simple solution to a problem is found, the design team should be challenged to either beat the competitor's solution or use it*. Many designers, when faced with this challenge, find amazingly creative and innovative ways to improve on their competitor's design. When stretch objectives are set in this way, probabilities become high that the resulting product will be superior to the competition in the areas where the stretch objectives are achieved.

Standardization and rationalization

Standardization and rationalization (S&R) is an approach which seeks to eliminate complexity and control proliferation of information throughout the manufacturing system. In the S&R approach, standardization is the reduction in the number of options (for example, parts, processes, and so forth) used in *existing* designs. Rationalization is the identification of the fewest number of options to be used in *future* designs. Black and Decker has used the S&R approach to greatly reduce the number of hardware components purchased by the com-

---

*This advice is offered by Don Clausing of MIT and others who advocate competitive benchmarking as a powerful and effective design practice.

pany (Bradyhouse, 1987). For example, the number of plain washers to be used in future products has been reduced from 448 to 7 (one material, one finish, one thickness). Similarly, the list of 266 ball bearings currently used has been rationalized to 12 (one seal, one lubricant, one clearance, metric only).

The rationalized options are used only in new designs, no attempt is made to retrofit existing products with the rationalized components. However, where possible, low-cost changes are made to existing product to eliminate unpopular options and increase the concentration of popular options. For example, if a company is currently buying 300 different cold-headed fasteners and 15 of these account for more than 50 percent of total fastener usage, then simple changes are made to increase the use of the 15 and eliminate the others. A rationalized list of fasteners may or may not include one of the currently popular fasteners; ideally, it will be considerably shorter than 15. The idea is to use standardization to simplify existing product and rationalization to simplify new designs.

The S&R principles can be applied to all aspects of the manufacturing system, such as parts and purchased components, materials, processes and methods, tools and fixtures, process operations, material handling methods, and tools for servicing products, to name just a few. To be effective, S&R is best implemented on a corporate- or business-wide basis. Standardized and rationalized lists of options should be developed by cross-functional teams and accepted, understood, and used at all levels of the organization. Therefore, S&R requires extensive planning, management commitment, and training. Also, S&R is greatly facilitated by the availability of an activity–transaction-based standard cost system.

Using S&R results in fewer of everything—parts, processes, tools, specials, job descriptions—and the accompanying reduction in manufacturing information content and complexity. Other major benefits include less design time, tested and proven options, established suppliers, favorable prices, and interchangeable parts for ease of service.

Process-driven design

In process-driven design, the manufacturing process plan is developed prior to performing the product design. Although having manufacturing-led product design represents a radical departure from the conventional product development approach, it can be a very effective strategy because it is based on the recognition that product design decisions often inadvertently limit the manufacturing options available for production of the product. This is especially true if advanced manufacturing technologies such as robotics, machine vision, and flexible manufacturing methods are to be used. Process-driven design keeps the product design from unnecessarily constraining manufacturing by providing up-front guidance to the design team before the design concept is frozen and allows both to converge in a uniform and controlled fashion.

Process-driven design is implemented by specifying process requirements and the preferred methods of manufacture as design requirements before design of the product begins. The product is then designed so that it can be manufactured in this most desirable way. In one approach, required end results are defined first in the form of goals for quality, reliability, productivity, and cost (see Fig. 4.8). Manufacturing then defines the best processes for building the product that will achieve these goals. Next, guidelines based on these "best" processes are developed to help ensure that the design features required by these processes are incorporated into the design. The product is then designed using these guidelines to take advantage of the best processes. For best results, the product design should be done in a synergistic team approach with manufacturing to ensure a coordinated final product and process design.

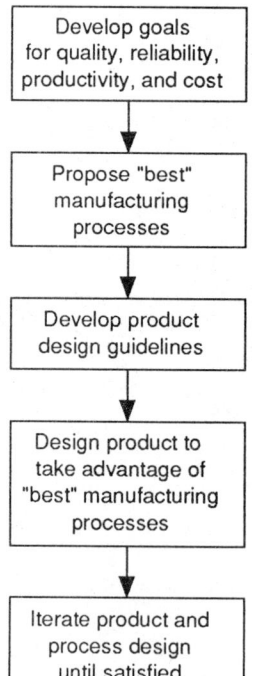

Figure 4.8 An approach for process-driven design.

Design for quality

Variability is the enemy of manufacturing. It is a major cause of poor quality resulting in unnecessary manufacturing cost, product unreliability, and ultimately, customer dissatisfaction and loss of market share. Variability reduction and robustness against variation of hard-to-control factors are therefore recognized as being of paramount importance in the quest for high quality products. In a design-for-quality approach, the design team seeks to design the product and process in such a way that variation in the product's functional characteristics due to variation in hard-to-control manufacturing and operational parameters is minimal. The ideas behind this approach are largely attributable to the efforts of Dr. Genichi Taguchi and the cost-saving approaches to quality control pioneered in Japan. An important element in this approach is the extensive and innovative use of statistically designed experiments.

The key to minimizing variability in a product's functional characteristics is to systematically select values for controllable factors such that sensitivity to uncontrollable factors is minimized (see Fig. 4.9). Controllable factors include all product design parameters (for example, material, dimensions, part geometry, design configuration), all

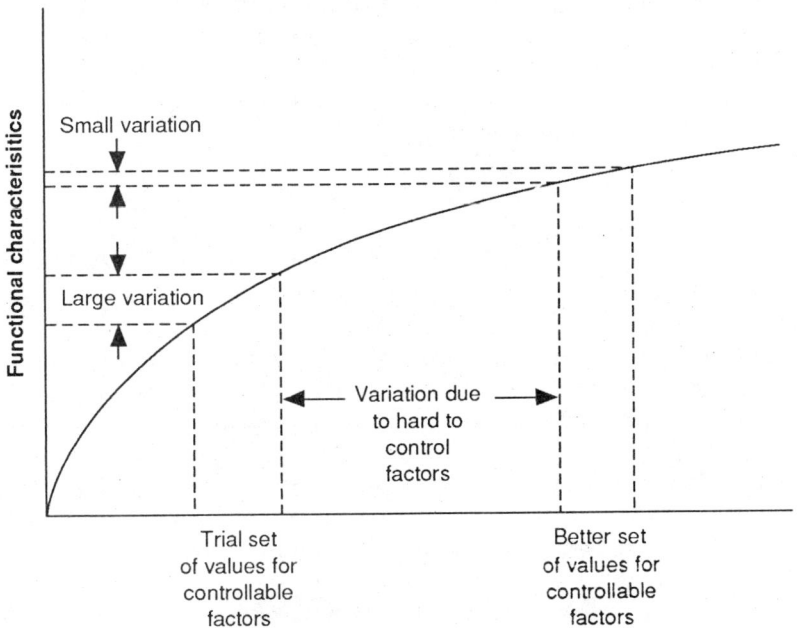

Figure 4.9 Parameter design exploits nonlinearities between product functional characteristics and product and process design parameters.

process design parameters, and all process setting parameters. Hard-to-control factors, called noise by Taguchi, include environmental factors, such as temperature and humidity; time and use factors, such as deterioration of materials and wear; and manufacturing related inconsistencies such as part-to-part variation.

As practiced by Taguchi (1979), design for quality involves a three-step optimization of product and process: system design, parameter design, and tolerance design. System design involves wise selection of materials, part configurations and geometries, tentative product parameter values, production equipment, and tentative values for process parameters.

In parameter design, tentative nominal values are tested over specified ranges to determine the best combination of parameter values. In the Taguchi method, parameter design is typically done by using fractional factorial designs and other orthogonal arrays in novel ways. If the reduced variation obtained by parameter design is not sufficient, tolerances on influential product and process parameters are tightened in the tolerance design step.

Tolerance design usually means spending money on higher precision, better grade materials, and more complex machinery. Many design and manufacturing organizations are conditioned to spend money to achieve required product functionality and performance. The tendency is therefore to jump from system design to tolerance design. In doing so, they omit the parameter design step where there is most to gain in terms of cost and quality. The key significance of the Taguchi method is its focus on this important step.

From the perspective of DFM, however, system design offers an equal or even greater opportunity for designed-in quality improvement. This is because it is in system design that the potential beneficial impact of design integration and the team approach is greatest. Often, inherent insensitivity to variation can be designed into the product and process during the system design step through creative application of the principle of independence of functional requirements (Axiom 1). Another approach in the system design step is to eliminate or minimize the source or cause of variation by design. Reducing the likelihood of assembly error by providing self-fixturing features that help orient and align parts for assembly is an example.

Design for quality can also be implemented in the system design step by intentionally designing the product and process to be tolerant of variation. Providing an easily recognized feature to ease factory floor lighting problems associated with vision system applications is a simple illustration of this approach. Here, a designed-in part feature essentially renders the vision system immune to variations in factory floor lighting, a process parameter that is traditionally difficult to control. Similarly, use of generous tapers and other guiding features makes part insertion less sensitive to another hard-to-control process parameter—assembly robot placement accuracy. Using a spring loaded support in place of a mechanical adjustment not only eliminates a potentially hard-to-control assembly and product operation parameter, but also helps make the product robust against deterioration over time and with use. Manufacturing complexity and cost are reduced, product maintenance is simplified, and the customer perceives the product to be of higher quality because no perplexing adjustments need to be worried about.

Management's challenge in doing design for quality in the system and parameter design steps is to create and nurture the necessary team approach environment needed to make it work and to provide the training and expertise required to implement it. Most importantly, management must expect design for quality, encourage it by providing opportunity and appropriate resources in the early stages of design, and act decisively on findings and recommendations. Bendell et al. make this point very well in the introduction to their book on world industry applications of the Taguchi Methods (Bendell, 1989).*

> Unless the findings are made known to those in authority and acted upon, there will be no savings, no production benefits, no productivity increase. Without decisions and subsequent action all the work is totally wasted.

Design for change

Change has become an increasingly important consideration in a product's life cycle. Customer needs and perceptions change, new product innovations and technology breakthroughs occur regularly, competition is constantly challenging and pushing current products, and new materials and processes are continually emerging. Change often initiates a design cycle, it occurs during the design cycle because of iteration and continuing uncertainties, and may well occur as the direct result of a new product introduction. Like variation, change creates chaos and noncompetitiveness when improperly managed or inadvertently overlooked. In a design-for-change approach, the design team seeks to make the inevitability of change compatible with the need for a stable and disruption free manufacturing environment.

In design for change, the objective is minimal capital investment and timing consequences incurred due to inevitable and necessary design change. Hard-to-control factors include:

---

*Bendall, p. 11.

1. Changing day-to-day production, customer, and market needs
2. Competitive pressure
3. Availability of new technologies and materials
4. Design iteration due to uncertainties
5. Design iteration due to continuous improvement in product and process.

Several strategies for designing a product and process to be robust against change appear to be possible. One of the most effective of these is standardization. The principle of modular design, for example, is a long standing design-for-change technique. The key in using standardization effectively is to identify the right standardization principle. Agreeing early to design a machined part so that all features can be obtained by machining on three orthogonal axis makes the design readily adaptable for manufacture on a flexible machining center. Use of a standardized base component for all variants of a particular product effectively decouples the material handling system from the particular product variant being assembled in a flexible assembly process. Reverse-engineering of many successful integrated product and process designs will reveal similar underlying standardization principles which, when adhered to by the product design team, enable the flexible manufacturing philosophy to be implemented.

In most designs, the scope of design or range of possibilities that must be considered typically diminishes as uncertainty is reduced. However, at some point in the design (usually after consideration has narrowed to one particular solution), new needs and opportunities begin to emerge. This broadens the scope of design to include many new possibilities and generally triggers a variety of redesign activities (Fig. 4.10). Recognizing the narrowing and broadening of design scope as a natural tendency of product development, we see that the best time to plan for redesign is during the early stages of the original design. By looking five or six product generations into the future, the design team can anticipate changes that are likely to occur and then consciously plan for them in the design. This can greatly reduce the design time and cost of new model introductions.

The following simple three-step procedure is suggested as a general approach for implementing design for change:

1. Evaluate the proposed product concept and process plan with respect to the following principle questions:
   - *How might the product change over time? How might customer needs or functional requirements change? What applicable new*

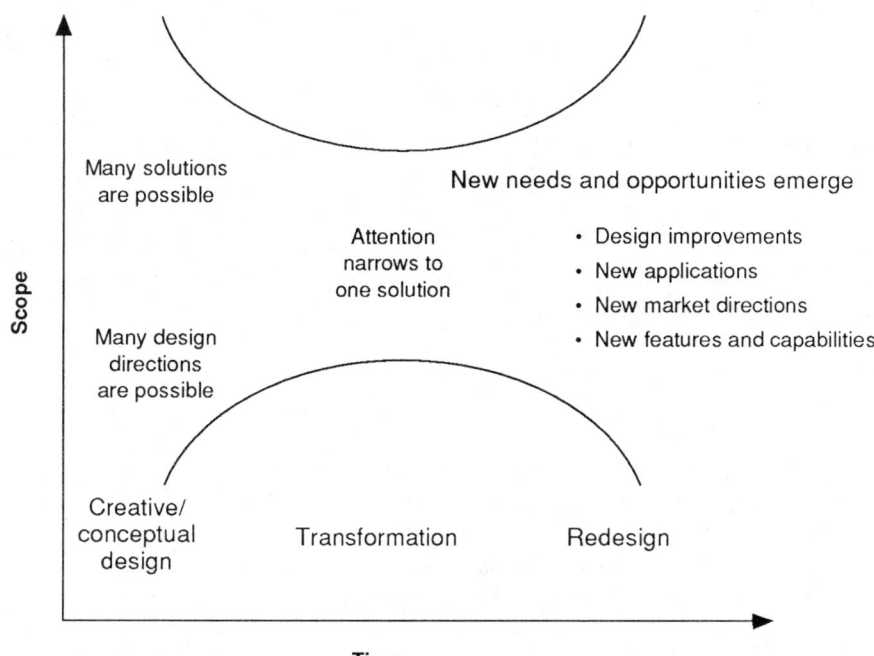

Figure 4.10 Scope of design as a function of time into the design.

technologies are likely to become available? How would the product be effected by these changes or new developments?
- *How might the process plan or production technology change over time. What effect would these changes have on the product design?*
- *What product or model variations are planned? How does the product-process concept accommodate these variations? What new variations could be introduced in the future? How would these changes impact the product design and process plan?*

2. Analyze the results of the evaluation and develop ideas and approaches in the design for accommodating expected change. For example, seek a systematized product structure that will facilitate introduction of changes that are deemed likely to occur and that will provide an easy migration path to future product generations with a minimum manufacturing impact. Seek integrated product and process concepts that decouple product change from method of manufacture. Divide the design into stable "chunks" that can be combined in different ways to produce a variety of different products within a defined product family using essentially the same production facilities and tools.

3. Improve the product design and process plan according to the ideas adopted and re-analyze. Iterate until satisfied.

In following this procedure, each question in Step 1 should be answered in as much detail as possible, given the product and process definition available. If the product design and process plan are fairly complete, then the procedure will help identify vulnerabilities to change. If the product and process concept is less well developed, then the procedure will help define change related design objectives and delineate concerns.

Design for flexible manufacture

The significance of flexible manufacture is the ability it offers to manufacture to customer order, manufacture a family of different parts or products in any sequence and quantity, and rapidly introduce a new product or make an engineering change to an existing product. *Flexible manufacturing is therefore, a major design-for-change strategy.* As companies gain experience with the DFM philosophy and the team approach, they often discover new and innovative opportunities for flexible manufacture as illustrated by the following examples.

Bulkhead subassembly is a problem in airframe manufacture because each bulkhead is a different size and may have a different stiffener configuration. Hence, different but essentially similar tooling is required for each bulkhead assembly. One U.S. aircraft manufacturer has solved this problem by using a reconfigurable tooling design. In this approach, the same tooling is used to assemble a variety of bulkhead subassemblies simply by reconfiguring certain clamping bars that adapt the fixture to different size bulkheads. However, to be usable, details of the bulkhead subassembly, such as tooling hole location and stiffener positioning must be carefully coordinated with the reconfigurable fixturing geometry. This coordination is currently done through close cooperation between design and manufacturing in a team approach.

As an alternative to using the team approach for each detail of bulkhead design, making the fixturing geometry available in the CAD environment would enable the design engineer to coordinate the fixturing configuration and bulkhead subassembly design as part of detail design of the bulkhead. Some companies have already done this type of CAD-based product and process integration successfully. The flexible tooling approach developed by a manufacturer of electric motors for automotive applications is an illustrative example of this.

In this particular case, the electric motor supplier was faced with frequent introductions of new electric motor designs to meet the needs except for their housing and mounting geometries. However, because the housings are different, new fixturing must be designed and built for each new motor design. In addition to the expense and lead time involved, the fixtures often require rework adding additional time and complication to production start-up.

To address this problem, a team of product design and tooling engineers developed a set of "fixturing primitives" that could be assembled together in a variety of configurations on a standardized base plate to meet the fixturing needs of all conceivable electric motor assemblies. To facilitate design of the fixture configurations, the team created a CAD library of the fixturing primitives. By using these CAD primitives, together with a variety of pop-up help panels providing design suggestions and guidance, the product designer is able to design the fixture as part of the new motor housing design. Printed assembly instructions for the fixture are released as part of the design release. With this approach, validated fixturing that works right the first time is available well ahead of schedule. This flexible fixturing approach not only reduces cost and design time, but also significantly improves on-time performance of the supplier.

In many mature product technologies, much of the cost reduction "cream" has been skimmed through on-going design optimization and producibility refinements. These examples demonstrate how the DFM philosophy can open up new avenues of opportunity for designing and manufacturing smarter through coordinated product and process design.

Design for analysis

By superimposing manufacturing and other life-cycle process requirements on the functional requirements, many choices are quickly narrowed to the best few. Suri and Shimizu have noted that essentially the same beneficial narrowing and focusing of choices occurs when simple, easy-to-use and easy-to-understand design/analysis tools are employed in the design process (Suri, 1989). This has led them to suggest a new strategy to improve the design process which they call Design for Analysis (D/A). In the context of the D/A hypothesis, analysis is defined as any procedure that ascertains whether a given design will meet certain specified objectives. A design tool is any means that assists the designer in performing design and analysis activities. The D/A principle states that "designers should be constrained to work with only those designs (of products and systems) that can be analyzed easily and quickly by simple tools."

Suri and Shimizu show, through the analysis of several design case studies involving the use of simplified tools, that these conjectures are supported through actual practice. Their findings suggest that limitations of the analysis tool capability can actually enhance design creativity rather than limit it. They also suggest that simplifying assumptions required to use simple tools leads to simple solutions. Also, in some cases, the use of simple tools work "as guide rails preventing the design from deviating from the design objectives."

Design for assembly

The Design for Assembly (DFA) method was developed by G. Boothroyd and P. Dewhurst while at the University of Massachusetts (Amherst). Details of the methodology are presented in *Product Design for Assembly* (Boothroyd, 1987).

Based largely on industrial engineering time study methods, the DFA method developed by Boothroyd and Dewhurst seeks to minimize cost of assembly within constraints imposed by other design requirements. This is done by first reducing the number of parts and then ensuring that the remaining parts are easy to assemble. Essentially, the method is a systematic, step-by-step implementation of aspects of the eliminate, simplify, and standardize approach discussed earlier.

For many companies, DFA has been the starting point for development of a corporate DFM philosophy and the culture change that accompanies it. Before DFA, it was taken for granted that the design engineer knew best how to design products. The job of the assembly engineer was to find the best way to assemble the product once the product was designed. Use of the DFA method quickly creates company-wide recognition of how problems caused by the serial approach can be remedied by shifting to a DFM philosophy and team approach. On-going experience with DFA has supported this recognition by irrefutably and consistently showing that designing a product, while keeping in mind how it will be assembled, offers significant benefits, including lower cost and higher quality. For companies wishing to get started with DFM, training in the DFA method followed by subsequent implementation in a suitable new product program and on-going nurturing by management is an excellent, expedient way to begin.

DFx tools

Design for "x" (DFx) tools help to ensure that parts and products are correctly designed to be produced using a particular production process or method such as plastic injection molding or sheet metal stamping. Design requirements for a given process are often stated in the form of design guidelines and rules of thumb. Typically, these guidelines are highly specialized for a particular industry, process implementation, plant, or equipment installation within a particular plant. DFx tools help the designer to systematically consider these process requirements and constraints early in the design process. Examples of DFx tools include design for casting, design for injection molding, and design for metal stamping. The "flexible tooling" design tool for automotive electric motors discussed previously is another example.

A major goal in the development of DFx tools should be the prevention of poor part designs from being released to manufacturing. One way to prevent poor designs from being released is to include a producibility rating or score as part of the DFx tool. Acceptable designs would receive scores greater than some minimum value established by the design team. Best designs for manufacture would receive a maximized score. To be effective, the DFx tool should also point out problems and suggest alternative fixes when low scores are achieved.

DFx tools augment the team approach. They help in situations where geographical distance between design and manufacturing may make the team approach difficult or where there is a shortage of experienced manufacturing engineers. They provide constraints that lead to shorter design times. They teach good design-for-manufacturability practice, helping young and inexperienced designers come up to speed quickly and encourage the creative designer to find innovative ways of achieving producibility as an integral part of performing the functional design.

Producibility measurement

The ability to measure producibility is a major DFM objective. Once producibility can be measured and expressed in numbers, then something concrete is known about it and positive actions can be taken to improve it. Producibility measurement makes it possible to implement DFM on a more quantitative basis.

At present, producibility measurement is in its infancy. Concepts such as assembly design efficiency and other producibility ratings are beginning to be used commonly, but no macroscopic, all encompassing producibility measures have been agreed upon. This is largely because effective producibility measurement must include the integrated effects of many separate as well as coupled behaviors related to product, process, and material.

Some producibility elements which need to be measured include robustness, rolled yield (cumulative probability of conformance with respect to product, process, material, and component parameters), cost, complexity, assembly ease, cycle time, process risk, material risk, and environmental risk. A seemingly endless number of producibility measures can be proposed for most of these elements. Complexity, for example, is measured by the number of functional requirements to be satisfied, total number of parts, number and variety of manufacturing processes, number of nonstandard or special or unusual or unique conditions or requirements, number and variety of tools required, touch labor hours for assembly, component layout and part geometry, number of inspection points needed, and estimated Mean Time To Repair (MTTR) to name just a few. In selecting producibility measures, it seems advisable to narrow the list to the one or two most important for a given element. Hence, *total number of parts might be selected as the one best measure of complexity*.

A company's ability to define and measure producibility, as well as its willingness to do so, will increase greatly as the DFM philosophy becomes more ingrained and institutionalized. Development of usable producibility measures can offer a considerable competitive edge. A producibility measurement approach to DFM might consist of the following steps:

1. Define appropriate quantitative producibility measures.
2. Capture necessary process, material, and component data across all products currently in production and summarize into indices of capability that benchmark the company's overall design and manufacture ability.
3. Carefully analyze such benchmarking information to discover how to capitalize on the company's strengths and overcome areas of limitation.
4. Develop design methods which can be used during the design process to optimize a design with respect to agreed upon producibility metrics and allow producibility to be traded-off against major program drivers such as performance, reliability, and maintainability.

DFM toolkit

As a company adopts and institutionalizes the DFM philosophy, design and analysis tools that were originally used independently by various functions become tools for use in the concurrent engineering environment. The synergism that exists under these new conditions can change the way traditional design and analysis methods are used, provide new application dimensions, and leverage them in ways not before considered. For example, value engineering has traditionally been practiced late in a design project after many irreversible decisions have been made. When used on a continuous basis from the be-

TABLE 4.1 Design and Analysis Tools and Concepts for DFM

DFM tool	Description	Remarks
Design for Assembly	Systematic method for simplifying a design by reducing the number of parts and ensuring that the remaining parts are easy to assemble.	Rapidly effective and easy to learn. DFA is easiest starting point for DFM.
Computer-Aided-DFM	Computer-based tools which help integrate product and process. Includes a variety of tools ranging from variation simulation analysis to solid modeling and design with feature techniques.	Saves time and can simplify effort. Facilitates "what-if" optimization. Fosters team building.
Quality Function Deployment	A new method providing for the translation of the assessed customer needs into technical requirements for each stage of the production process.	Provides a traceable link to customer requirements. Provides a practical base for continual improvement.
Taguchi method	Seeks to define a robust combination of design parameter values through use of fractional factorial designs and orthogonal arrays.	Based on powerful quality engineering concepts. Moves quality into design stage.
Statistical problem solving	The simultaneous study of many factors and their interactions.	These methods help to build good quality into products.
Statistical process control	Use of statistical monitoring and control techniques to achieve desired outgoing quality in products.	Helps to surface product design problems and to guide tolerancing and other product design practices.
Group technology	A technique for exploiting the sameness or similarity of parts based on their geometrical shape and similarities in their production process.	Facilitates standardization and rationalization. Attacks part proliferation and saves design time.
Failure mode and effect analysis	A methodical way of studying the cause and effects of failures before the design is finalized.	Helps prevent failures and defects from occurring and reaching the customer.
Value engineering	Systematic application of recognized techniques which identify function, establish value for the function, and provide the necessary function at the lowest overall cost.	Offers an organized approach to assessment of cost impact of design decisions.
Activity-transaction-based accounting	Cost systems that allocate overhead cost to particular products and components based on activity measures.	Links design with "hidden" cost of doing business. Exposes and helps prioritize cost reduction opportunities.

ginning of an integrated product and process design by a cross-functional design team, value engineering becomes a totally different tool. Table 4.1 provides a list of traditional and not so traditional design and analysis tools and concepts that should be part of every design team's DFM toolkit.

Summary

DFM is recognizing the importance of integrated product and process design and then consciously going about design in a way that leads to a product inherently easy to manufacture and support. By superimposing life-cycle process requirements on the functional requirements of a design, many choices are narrowed to a few good choices. Using a cross-functional team approach accelerates the rate of product knowledge growth and greatly expands the product knowledge base available for guiding and validating product and process design decisions. The result is a shortened time to market with a higher quality, lower cost product (see Fig. 4.11). Both the plateau that typically develops because of the integration of various sub-solutions and adjustments to satisfy manufacturing requirements, as well as the oscillations that occur at the end of the design cycle caused by manufacturing and assembly difficulties discovered late in the project, are avoided.

Providing *disciplined anticipation* that the product will be correctly designed for manufacture is the key action required by management in developing a good DFM program. Training the design team in DFM techniques such as Design for Assembly, as well as

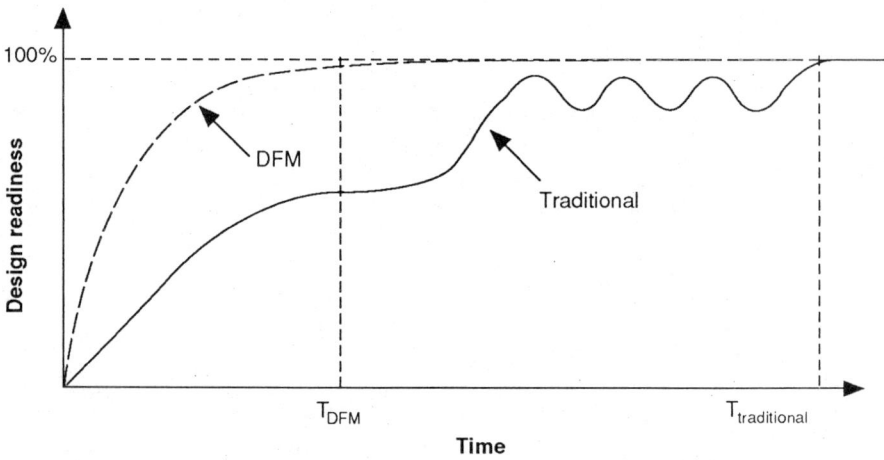

Figure 4.11 Time history of design.

providing a well integrated kit of DFM tools, creates awareness and sets the stage. A well thought out design process enables DFM approaches to be implemented. Setting stretch goals stimulates innovation and sends a clear message that DFM is expected. Providing an adequate budget of time and money for up-front activities, together with a design environment that facilitates easy communication, encourages and makes a well planned total design possible. Personal commitment and interest in the DFM aspects of the design on the part of management make DFM an important part of the design project and sustains the expectation.

References

Bendell, A., Disney, J., and Pridmore, W., *Taguchi Methods: Applications in World Industry,* Springer-Verlag, 1989.

Boothroyd G. and Dewhurst, P., *Product Design for Assembly,* Boothroyd and Dewhurst, Inc., Wakefield, R.I., 1987.

Bradyhouse, R., "The Rush for New Products Versus Quality Designs That Are Producible: Are These Objectives Compatible?," presented at the SME Simultaneous Engineering Conference, June 1, 1987, Society of Manufacturing Engineers, Dearborn, Mich.

Bradyhouse, R., "Starting Up a DFA Program and Cashing in on the Opportunities of the Persistent," presented at the SME Simultaneous Engineering Conference, Society of Manufacturing Engineers, Southfield, Mich., November 9, 1988.

Dixon, J. and Simmons, M., "Expert Systems for Mechanical Design: V-Belt Drive Design as an Example of the Redesign Architecture," *Proceedings ASME CIME Conference,* Las Vegas, Nev., August 1984.

Guideline VDI 2221, "Systematic Approach to the Design of Technical Systems and Products," VDI Society for Product Development, Design, and Marketing, VDI-Verlag GmbH, D-4000 Dusseldorf, August 1987.

Huthwaite, B., "Design for Competitiveness," Bart Huthwaite Workshops, Troy Engineering, Rochester, Mich., 1988.

Huthwaite, B., "The Link Between Design and Activity-Based Accounting," *Manufacturing Systems,* vol. 7, no. 10, October 1989.

Rinderle, J., "Design of Design Automation Systems," *CIT Engineering News,* Carnegie Mellon University, vol. 6, no. 1, Spring 1986.

Suh, N., Bell, A., and Gossard, D., "On an Axiomatic Approach to Manufacturing and Manufacturing Systems," *Journal of Engineering for Industry,* ASME, vol. 100, no. 2, May 1977, pp. 127–130.

Suri, R. and Shimizu, M., "Design for Analysis: A New Strategy to Improve the Design Process," Technical Report No. 89-3, Dept. of Industrial Engineering, University of Wisconsin, Madison, Wisc., April 25, 1989.

Taguchi, G. and Yuin, W., *Introduction to Off-Line Quality Control,* Central Japan Quality Control Association, Nagaya, Japan, 1979.

Yasuhara, M. and Suh, N., "A Quantitative Analysis of Design Based on the Axiomatic Approach," in *Computer Applications in Manufacturing Systems,* ASME Production Engineering Division Publication, PED-2, 1980.

Part 2
Case Histories

Chapter

5

Revitalizing the Manufacture and Design of Mature Global Products

Alvin P. Lehnerd*

Introduction

Manufacturing enterprises are evolutionary entities. Over time, their product portfolios expand through evolutionary and chronological developments. Products are usually designed and developed one at a time. As a result, it is the exception when the designs of a manufacturer's products embrace much compatibility, standardization, or modularization. The norm is that product portfolios are rarely designed simultaneously; designs take place in a sequential manner. Additionally, many current products of U.S. manufacturers were designed and tooled years ago, yet prevailing labor rates, manufacturing processes, energy costs, availability of materials, and interest rates are often significantly changed from the time of the original product design and tooling activities. It is rare that a U.S. manufacturer invests the time and resources necessary to rationalize production of an entire product line to fit the changing economic environment and to take advantage of opportunities provided by technological advance.

Manufacturers usually design for function, then redesign for manufacturing; thus, two design iterations usually take place. If an enterprise wishes to maintain or gain market share in global markets, the firm's managers and technical personnel must learn to combine man-

---

*This chapter originally appeared in the book *Technology and Global Industry,* National Academy of Science, National Academy Press, pp. 49–64, 1987. It is reprinted here by permission with slight changes to fit the format of this book.

ufacturing with innovation in product design. Few enlightened companies take time for a third design iteration to automate and mechanize production for global leadership in cost and value.

In many, if not all, instances, design for manufacturing is also constrained by the existing resources of plant and equipment. In other words, manufacturing engineers guide the design decisions to match the profiles and capabilities of their existing factories and their respective in-house capabilities. In-place facilities are frequently barriers to product innovations. Fixed capital investments in existing capabilities are also barriers to more advanced lower-cost processes. Organizations commonly ignore what the production cost could be if their products were not shackled to outdated manufacturing processes and could also use state-of-the-art materials requiring new processes and procedures.

An additional issue is that few U.S. domestic manufacturers look at their product offerings as global opportunities. This domestic myopia—the belief that the marketplace ends at the U.S. borders—is a problem for U.S. industry, and the problem will only get worse as the world becomes more economically integrated.

Finally, corporate planning horizons are too short, and manufacturers seldom ask themselves what they are doing to ensure their longevity in the business. Too many managements or boards of directors do not act until external influences cause significant disruptions and spur the organization into action.

This chapter presents a case history of a 1970s program at Black & Decker Corporation to redesign a product line for production automation and leadership in cost and value. The program was an effort to redesign standard products to take advantage of opportunities for using new materials and new manufacturing and design techniques.

Black & Decker

When managers at Black & Decker Corporation observed growing global competition in the 1960s and 1970s, they decided that a window of opportunity existed to improve their product lines and manufacturing capability. Moreover, they decided that if they did not take time to do it right the first time, they would never have the time to do it over. They recognized that if they were to be a domestic manufacturer with aspirations to do business internationally 20 years hence, they would have to change the business in a way that would ensure that long-range performance. This involved making certain irrevocable decisions.

The impetus for change came from three sources. First, it was evident that foreign competition would increase in Black & Decker's product markets and that this would lead to foreign participation in new, related product markets.

Second, in the 1970s, inflation in costs of labor, material, services, and capital goods was a serious consideration. Table 5.1 shows the effect of inflation in the labor component of product costs. It assumes an 8 percent compounded inflation rate over five periods from year one to year six. To maintain constant labor-cost content in the product, one-third of the labor has to be removed from the product between period one and period five. In Black & Decker's assessment, offsetting inflation in labor costs depended on making better use of labor in adding value through design standardization, mechanization, automation, better use of material and floor space, and intelligent capital planning.

The third factor in Black & Decker's decisions was an anticipated continued public attention to consumer protection and environmental concerns. In the power tool industry, this attention took the form of requirements for double insulation of tools. The term "double insulation" refers to the additional insulation barrier placed in an electrical device to protect the user from electrical shock if the main insulation system ever fails. In the late 1960s there was a strong possibility that double insulation of domestic power tools would be legally required. Black & Decker decided that the threat of required double insulation provided an opportunity to study the entire product line.

The program begun at Black & Decker in the early 1970s was called Double Insulation. All consumer power tools were to be redesigned at the same time and would initially be manufactured in various locations in North America with standardized parts and components.

Double Insulation was Black & Decker's vehicle to redesign the line and develop a "family" look, simplify the product offering, reduce manufacturing costs, automate manufacturing, standardize components, incorporate new materials, improve product performance, incorporate new product features, and provide for worldwide product specifications. The program was designed to incorporate double insu-

TABLE 5.1 Impact of Wage Inflation on Labor Costs (8 percent compounded inflation)

Year	Hourly wage, $	Labor minute value of $3.00
1st	3.00	60.0
2d	3.24	55.5
3d	3.50	51.5
4th	3.78	47.6
5th	4.08	44.1
6th	4.41	40.8

lation on 122 basic tools with hundreds of variations. Of 18 tool groups manufactured by Black & Decker, 8 contributed 73 percent of sales, 71 percent of earnings, and 91 percent of units sold. These groups were tools and drivers, jig saws, shrub and hedge trimmers, hammers, circular saws, grinders and polishers, finishing sanders, and edgers.

Many of Black & Decker's U.S. competitors that had already introduced products with double insulation had incurred a 15 to 20 percent premium in material and labor costs in doing so. It was Black & Decker's goal to add double insulation without increasing the cost of any new tool beyond that of the existing product. In addition, of course, Black & Decker aimed to avoid dilution of assets or return on investment.

In this instance, Black & Decker's decision to introduce fundamental redesign throughout its product line was motivated by the prospect of an industry-wide requirement to incorporate double insulation in power tools. At other times, competitive product analysis plays an important role in decisions to redesign (the Appendix to this chapter describes a competitive product analysis carried out by the Sunbeam Appliance Company).

An important part of the plan for Double Insulation was the decision that the resources of the organization would be concentrated on this transition. Black & Decker would leave only a small portion of its management and engineering staff to carry out development efforts on new products. The development of new products was put in abeyance, and the resources usually devoted to development were focused on the manufacturing processes essential to the program.

To accomplish the engineering goals, a bridge was needed between design engineering and manufacturing. That bridge was the placement of advanced manufacturing engineers in residence at headquarters to work elbow-to-elbow with the design engineering groups (see Chaps. 1 and 12). These manufacturing engineers were involved with machine development, process development, value and cost engineering, purchasing engineering, and packaging. In addition to bringing manufacturing and design together at the engineering level, the basic structure of the company was changed. Before these changes were made, the program structure had consisted of a general manager and two vice presidents—one for manufacturing, and one for engineering and product development. That organization was changed, and a new position—vice president of operations—was developed to combine manufacturing, product development, and advanced manufacturing engineering under one manager.

A final general point about the Double Insulation program was the large investment required and the long time horizon needed to reap a return on that investment. As Fig. 5.1 shows, the break-even point in the program came nearly seven years after the program began, and

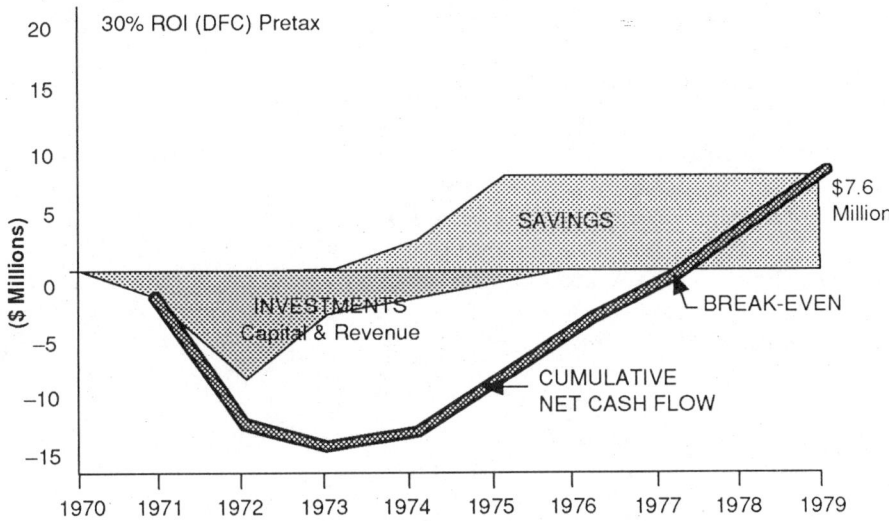

Figure 5.1 Financial analysis of Double Insulation program.

total cost was $17 million. Figure 5.2 shows the cumulative cost of the program from 1971 through 1975. Capital expenditures were $6 million. Tooling was $6.5 million. And development engineering and manufacturing technology were $1.7 million each. It is important to note that this program is rare from the standpoint that as much

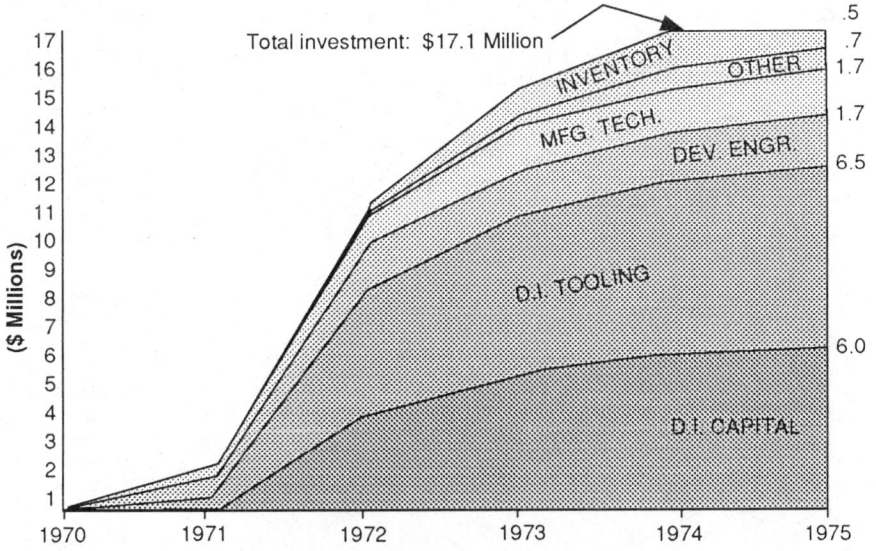

Figure 5.2 Investment requirements for Double Insulation program.

122   Case Histories

money was spent on manufacturing technology as on development engineering (see Chap. 3).

The transition to Black & Decker's leadership in cost and value was the result of collaborative effort among design, manufacturing, and manufacturing engineering functions. The changes in design and production of motors are one example of this collaborative effort.

The most common component in all power tools is the universal motor. Figure 5.3 shows all the components of such motors before redesign. Figure 5.4 shows the motor configuration both before and after redesign. Motors are now manufactured automatically, untouched by human hands. The laminations, placed at the head of the mechanized line, are stacked, welded, insulated, wound, varnished, terminated, and tested automatically. Table 5.2 shows, at 1974 volumes of 2,400 pieces per hour, that the new Double Insulation manufacturing system required 16 operators and that the old design would have required 108 operators. Material, labor, and overhead cost is $0.51 in the old system and $0.31 in the new. The labor content cost is $0.02 in the new system, down from $0.14 in the old.

Through attention to standardization, the entire range of Black & Decker power tools could be produced using a line of motors that vary only in stack length—that is, standardization froze the dimensional

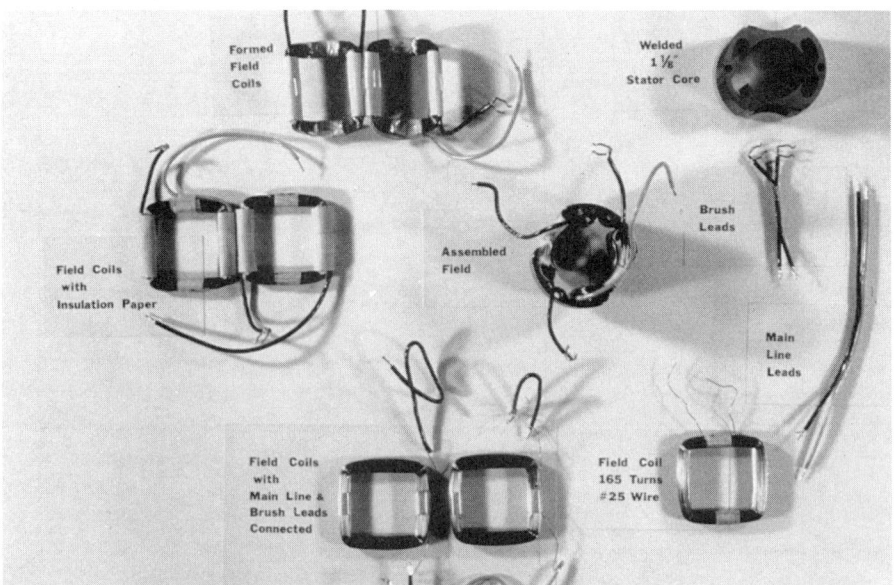

Figure 5.3   Electric motor field components.

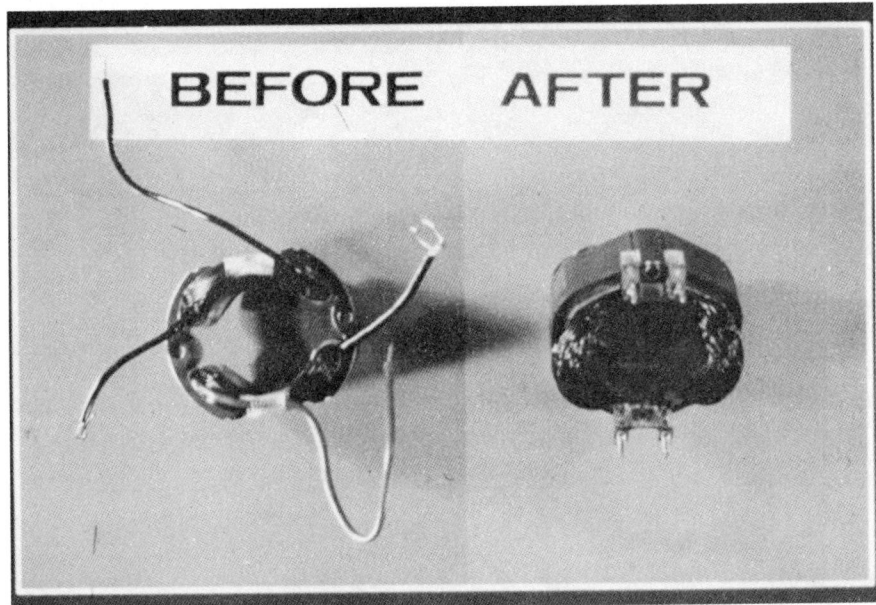

Figure 5.4 Motor configuration before and after redesign.

TABLE 5.2 Motor Field Production, Operator Requirements, and Costs at 2,400 Units per Hour, Old and New Design and Manufacturing Processes

	Today	Double insulation
Operators at 600/hr.	27	4
Operators to produce 1974 volume	108	16
Cost to insulate (materials, labor, overhead)	$0.51	$0.31
Labor Cost	0.14	0.02
Capital to produce 1974 volume	$400,000	$1,222,000
Annual savings (labor and materials only): $1,280,000		

geometry of the motors in the axial profile. All motors can now be produced on the same machines, and the only difference is stack length and the amount of copper and steel used. As Figure 5.5 shows, designs ranged from 60 watts to 650 watts, and the only dimension that changed was in the axial profile. The only difference in cost from the low-wattage to the high-wattage motors was the cost of materials and machine time; labor cost remained the same through the entire wattage range.

Another effect of design for manufacture can be seen in the produc-

124 Case Histories

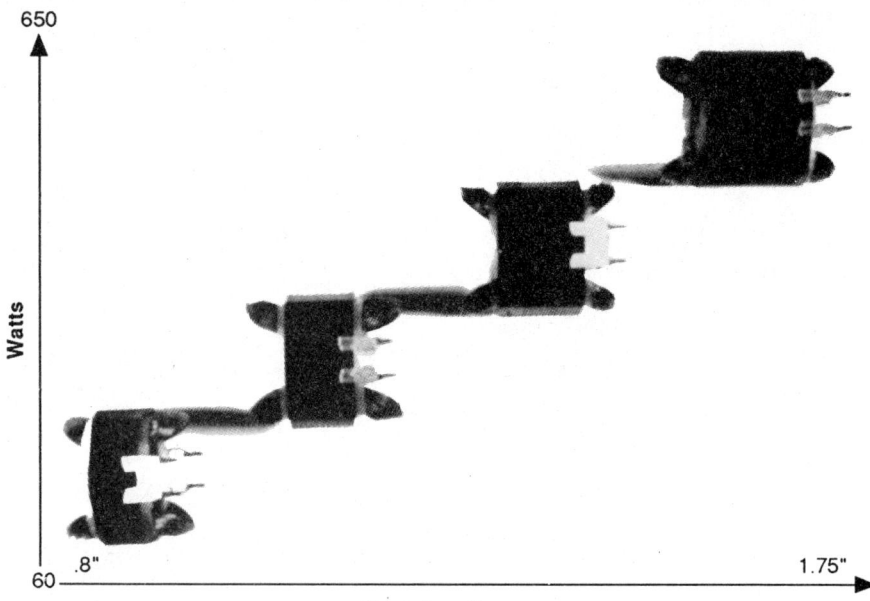

Figure 5.5 Motor stack length, 60 to 650 watts.

tion costs for the armature of the motors. As Table 5.3 shows, four times as many operators would have been needed to produce armatures under the old system as under the new system at a constant production volume. The effect on labor costs was dramatic. The labor cost of the manufacture using the new design was only $0.025 per unit, whereas the cost using the old design was $0.108 per unit.

The Results of Double Insulation

The Double Insulation program worked for Black & Decker. It reduced production costs, created opportunities for profitable vertical integration, increased market share, and improved the company's capability for new product development. Each of these changes is discussed further in the following sections.

Cost reductions

Cost reductions because of the Double Insulation program came mostly from labor savings, and the balance came from reduced factory overhead, material savings, and savings from standardization of parts

TABLE 5.3 Armature Production, Operation Requirements, and Costs at 1,800 Units per Hour, Old and New Design and Manufacturing Processes

	Old design and manufacturing process	New design and manufacturing process
Operators to produce	60	15
Cost to insulate (materials, labor, overhead)	$0.26	$0.11
Labor cost	$0.108	$0.025
Capital to produce	$2,340,000.	$795,000
Annual savings (labor and materials only): $540,000		

and modularization. In 1976 Black & Decker reviewed the program and compared existing equipment and labor costs with the capital equipment and labor costs that would have been required for the 1976 volume without the Double Insulation program (see Table 5.4). If the company had not carried out this program, estimated 1976 requirements for motor manufacture would have been nearly 600 people whereas the new system required only 171. That is a labor cost difference—from $6.4 million down to $1.8 million—of $4.6 million per year. The capital investment for the new system was higher than simple capital replacement—$4.6 million instead of $3.0 million—but with labor savings of $4.6 million per year, the payback on the investment was 4 months.

In its 1974 annual report, Black & Decker published its assessment of the effect of this project on four basic power tools (see Fig. 5.6). In current dollars, Black & Decker's power drills, for example, were $10 cheaper in 1973 than they were in 1958.

Figure 5.7 shows substantial reductions in the real cost of Black & Decker's products. The constant-dollar cost of products A, B, and C

TABLE 5.4 Motor Design, Old versus New Impact Analysis, Labor, and Capital Investment

(Including floor space, $20/ft^2)

	1972 Capacity		1976 Capacity	
	Old	New	Old	New
People	252	86	596	171
Annual labor	$2,700,000*	$ 900,000	$6,400,000†	$1,800,999
Capital	$1,300,000	$2,300,000	$3,000,000	$4,600,000

*Labor savings $1,800,000/year. Payback 1.25 years
†Labor savings $4,600,000/year. Payback 4 months.

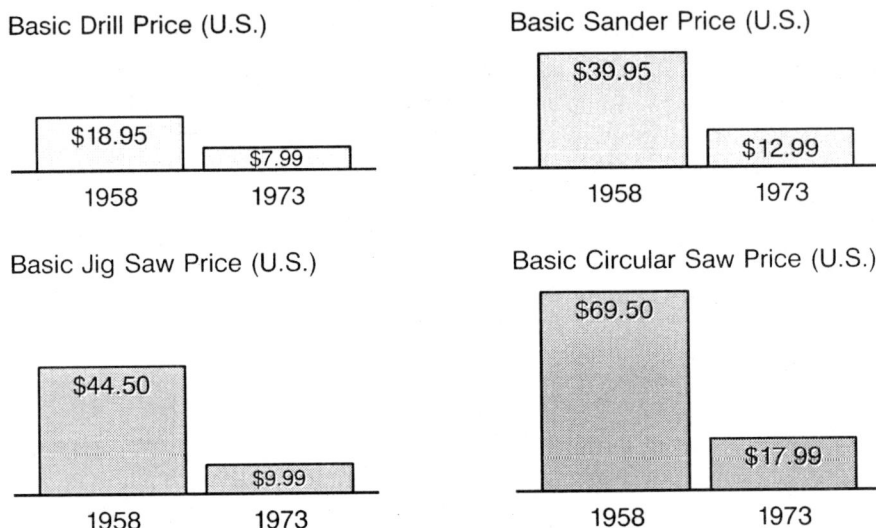
Figure 5.6 Prices, 1958 and 1973, of four basic hand power tools.

dropped by 47, 62, and 55 percent, respectively. The company was able to produce each product at an almost constant current dollar cost despite steady inflation in materials and labor costs. For Black & Decker's pricing position in the marketplace, the relevant comparison is between the top two lines on each graph, which show the difference, in current dollars, between manufacturing costs with and without Double Insulation.

Increased vertical integration

The cost and value leadership permitted unprecedented low prices to the consumers and thereby expanded Black & Decker's market share and increased household penetrations of power tools. The expanded volume resulted in opportunities for cost-effective vertical integrations. Examples include:

- Use of plastic materials grew from thousands of pounds per year to millions of pounds per year. Black & Decker's molding facilities were able to justify railcar bulk shipment of uncolored plastics resulting in a cost advantage of 5, 10, and sometimes 15 percent per pound. The coloring of plastic compounds at the molding machine reduced inventories, provided instant response to color changes, and eliminated material handling.
- Standardization of gears, and design revisions that allowed the change to spur gears from bevel gears, permitted the use of gears made from powdered metal. This change eliminated the need for gear cutting and

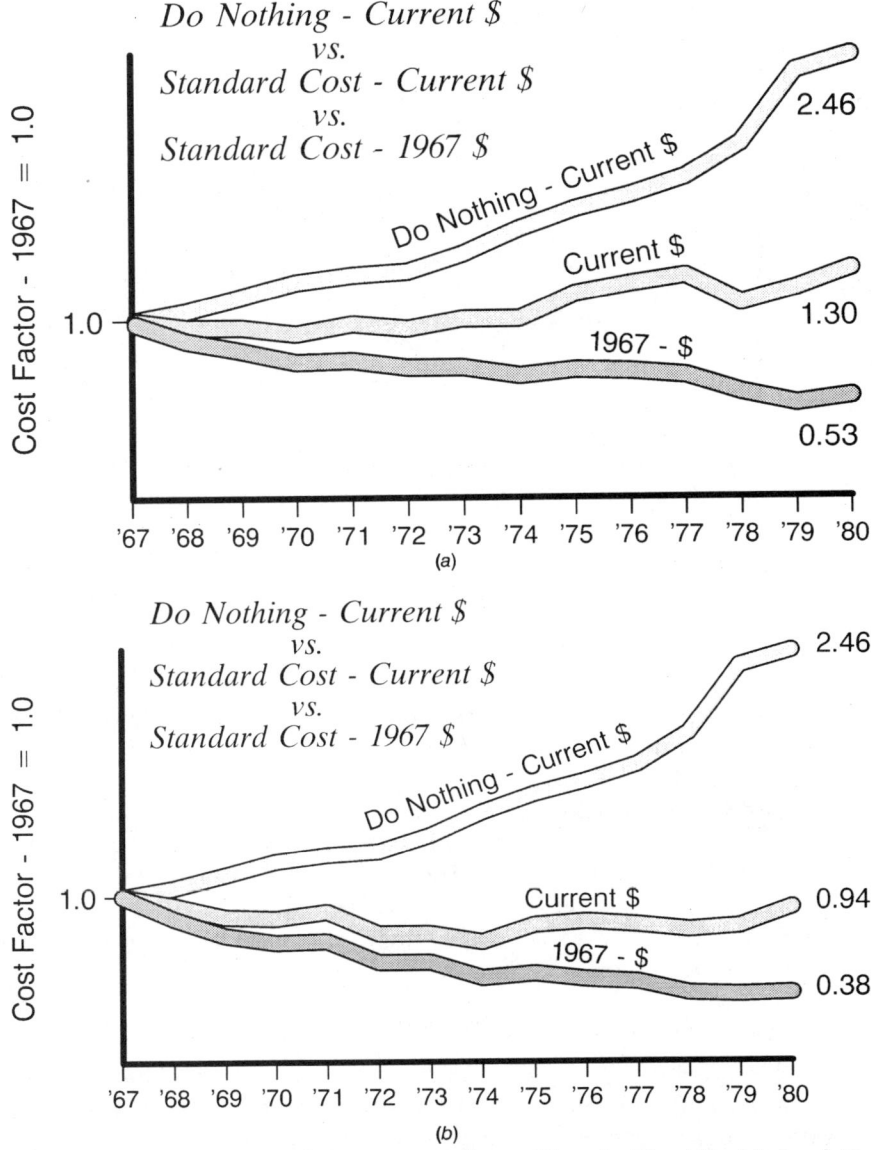

Figure 5.7 Product cost trends in current dollars with and without Double Insulation and product cost trend with Double Insulation in constant 1967 dollars for three products, 1967–1980. (*a*) product A; (*b*) product B; (*c*) product C (see the following page).

hobbing, heat treating, and gauging. These activities all contributed to high capital cost, high labor cost, and inefficient use of material in production. The volumes were large enough to permit vertical integration of fabrication of powdered metal gears.

- Before the Double Insulation program, 29 percent of the total cost of a drill was in the cost of a purchased chuck. Production volumes,

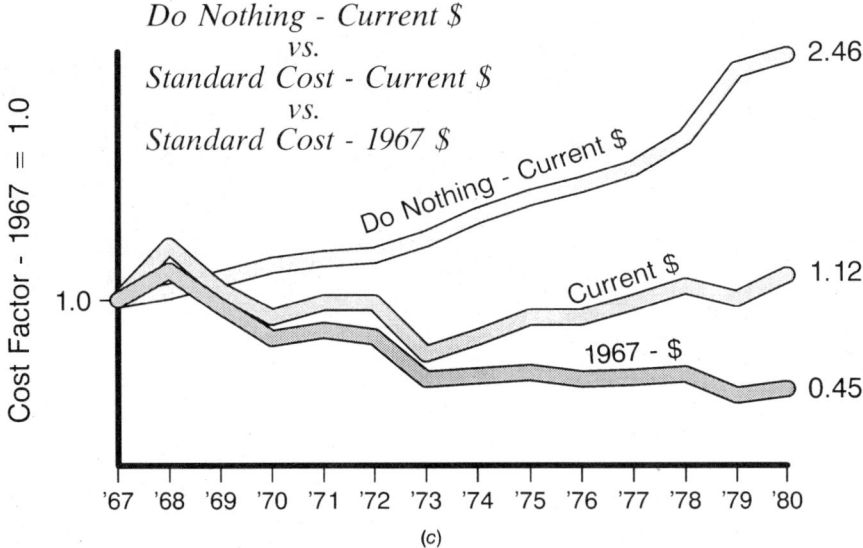

Figure 5.7 (*Continued*)

again, coupled with a modern state-of-the-art processing system enabled backward integration into chuck manufacturing at reduced costs of about 40 to 50 percent.

Standardization of bearings, switches, cord sets, cartons, fasteners, and so on, resulted in component volumes high enough to justify seeking sources on world markets for the best price.

The "inflation offset" idea proved to be recursive in that low production cost permitted low sales price, which increased unit sales, fueled vertical integration, and further reduced costs.

Competitive performance and market share

In the U.S. market, Black & Decker's competitors in consumer power tools were caught absolutely flat-footed. Their product designs and manufacturing processes were costly, and in an attempt to continue to compete they tried to match Black & Decker prices. This diluted their profitability and collapsed their ability to redesign to match Black & Decker. In the resulting shakeout in the market for consumer power tools, Stanley, Skil, Pet, McGraw Edison, Sunbeam, General Electric, Wen, Thor, Porter Cable, and Rockwell all left the consumer market. Sears Roebuck and Co. was able to stay in the domestic consumer market with Black & Decker.

In the European market, consumer power tools were much more expensive because the tool offerings were different. European tool producers provided a power driver in the drill configuration, but all other power tools—sander, circular saw, hedge trimmer—were sold as attachments. The availability of new low-cost single-purpose power tools enabled the consumer to eliminate the inconvenience and performance compromise of attachment technology. Black & Decker performed well throughout Europe as the new tools both greatly expanded Black & Decker's market share and increased household penetration.

Impact on new product development

Another benefit of Black & Decker's efforts was a substantially improved ability in product development. As new product concepts emerged, much of the work in design and tooling was eliminated because of the standardization of motors, bearings, switches, gears, cord sets, and fasteners. Design and tooling engineers working on a new product had only to concern themselves with the "business end" of the product and to perfect its intended function. New designs could be developed using components already standardized for manufacturability. The product did not have to start with a blank sheet of paper and be designed from scratch.

As products reached their maturity and had to be dropped, massive write-offs and scrapping of tools or equipment were necessary. This flexibility allowed marketers and managers to pivot quickly and avoid being tied to a dying product because they could not afford the write-offs.

In short, the pace of new product development and product retirement was greatly accelerated. Products could be introduced, exploited, and phased out with minimal expense related specifically to the decision to develop or retire a product.

Summary and Conclusions

In accomplishing the dramatic cost reductions through the Double Insulation program, the attitudes of the management were extremely important. Black & Decker management had a target of 15 percent compounded growth rate, and they wanted to remain independent and to service world markets. To do these things, the management focused not on marketing or financial manipulation, but on cost and value leadership in the industry. The management, as a team, projected how a successful program of this type could affect the marketplace and had the courage and tenacity to see it through. Black & Decker was also fortunate in having a large amount of latent talent. Many ordinary people proved capable of performing extraordinary tasks. With pricing and promotional strategies, the corporation was able to provide enough growth with the product improvements to avoid either reductions or expansions of its labor force.

Although the success that Black & Decker had with power tools may not be replicated in other industries, the principle of redesign and retooling for cost and value leadership in global markets can provide a focus for other manufacturing firms. It is a valid approach to achieving global competitiveness in manufacturing. By this means, U.S. firms can design, develop, and manufacture world-class products in the United States and at the same time achieve leadership in product value.

Appendix 5.1 Competitor Analysis by Sunbeam Appliance Company

In 1982 Sunbeam Appliance Company launched a program aimed at capturing at least 30 percent of the worldwide market share for the steam/dry iron in markets they wished to participate in. The first step was to assess global production capabilities and methods. Sunbeam obtained samples of competitive products from around the world for analysis of materials and labor content—estimated in time, not dollars. Components were reviewed and estimates of production costs were developed for all of the designs. After this material was pulled together, project management convened a two-day review for Sunbeam engineers from Australia, Germany, England, Canada, Mexico, and the United States to talk over what had been uncovered in this global product evaluation.

That analysis revealed some interesting aspects of the design and manufacture of steam/dry irons around the world. The number of parts used in the product ranged from a high of 147 parts to a low of 74 parts. The number of fasteners ranged from 30 to 16, and the number of fastener types in any one design ranged from 15 to 9. Sunbeam's existing product used 97 parts with 18 fasteners in 10 configurations. Reducing the cost of that design, incorporating everything learned from composite design, yielded a design which had 73 parts, 13 fasteners, and 7 types.

To gain a significant share of the market, however, it was necessary to leapfrog existing products and come up with a design with significantly lower cost and complexity over competitive offerings. Such a design was developed, with 51 parts and 3 fasteners in 2 configurations. Figure 5.A1 makes the point that driving down the part count also drives down cost.

Although reducing the part count entails a considerable effort in design engineering, effective design and process engineering will drive down labor and material costs.

The result of that effort was a composite design that would be the

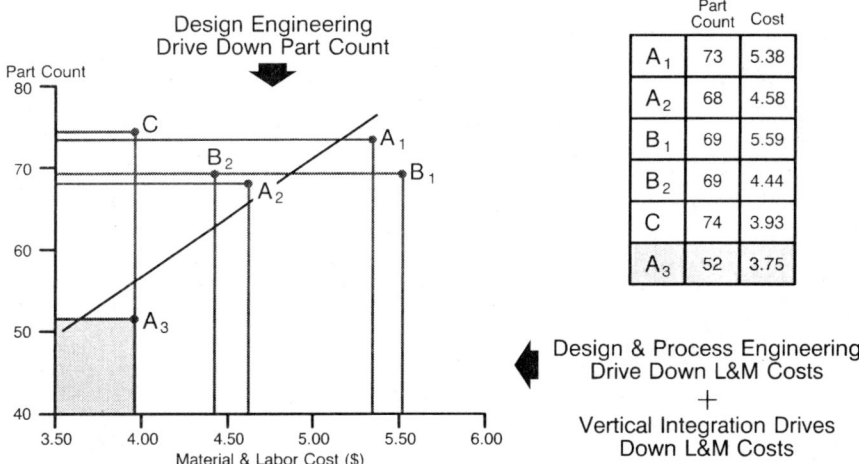

Figure 5.A1 Relationship of part count to material and labor cost per iron.

best of all of the products collected with attention to what the product would cost if the design were used throughout the world and compatibility were maintained. The new design is substantially cheaper to produce than either of Sunbeam's existing designs. The product was launched in 1986.

Chapter 6

GM: The Quad-4 Engine

John E. Ettlie, Stephen Nowak, and Klaus Blache

Introduction

The Quad-4 automobile engine was born out of necessity, after the oil shortages of the late 1970s, the increased pressure for lighter cars to meet the CAFE (Corporate Average Fuel Economy) regulations, and, perhaps most importantly, the need to revitalize a temporarily closed plant with a proud past to become competitive with a world-class product—the envy of the competition. The Delta Engine Plant, which makes the Quad-4 engine, produced a V-6 diesel for GM before the market evaporated for all diesel automobiles, especially those with the V-8. But the plant shutdown acted as a significant transitional event for the local GM organization in Lansing, Michigan. The diesel plant had never reached full production volume because of slow sales, and the employer had never been able to realize the full potential of plant efforts. This set the stage for the proposal of a new engine with a significantly different plant organization and set of processing principles and practices. Although it was an opportunity for GM to attain a leadership position as a high-technology engine producer the engine can best be characterized as a case study of "safe technology done right." The program product proposal drew on the strengths of an organization that had an unusually close working relationship between marketing and engineering, which were located together before and during product planning for the Quad-4. The central theme of this case history is the use of quality as the guiding principle in all plans and strategies for implementation (Miles, 1988). However, in the final analysis, the question that must be asked is "What price quality"? From the standpoint of market acceptance and satisfaction with a quality product, few cases are better examples than the Quad-4. Documentation of customer ratings, quality, and market impact are included in the last section of this case history. These numbers are impressive. The program single-handedly restored confidence in a threatened organization and provided a stage to launch other programs. However, most participants in the program would agree that it would take a GM or similarly large organization to duplicate this type and level of effort—completely new engine was launched from approval of project to production engines in 33 months. What is more, such production innovations as "no hot test" in the plant—the engines are first started when the cars (currently the Cutlass Calais, Buick Skylark and Pontiac Grand Am) roll off the assembly line—have actually resulted in fewer engine pulls (replacement at assembly).

Such marketing innovations of the program as an engine replacement policy for any engine problem during the first year of production led to the complete reverse-engineering of the GM's own product and subsequent improvements by the plant people that actually produced the Quad-4. It is likely that no engine has ever been tested more and pushed harder to guarantee reliability in the history of the domestic automobile industry. The high cost of the Quad-4 program was partially offset by the lessons learned and utilized throughout GM. This concerted program for corporate technology transfer is just one of the unique features of this simultaneous engineering case.

In the course of developing this case history we interviewed 14 members of the Quad-4 program and a representative of one of the major equipment suppliers for the production of the Quad-4 during the period from October 1988 to May 1989. We used existing records, articles, technical reports, and internal documents from GM, which were generously shared during the interviews.

The case illustrates three generic points that are central to the understanding of the management of design-manufacturing engineering. First, the relative dominance or importance of various functions in a cross-functional team approach to simultaneous engineering will depend to a great extent on the *degree of outsourcing* required for the product. This is not to say that the make-buy decisions are not revisited for an automobile engine that typically has a large outsourced set of special components. But, in the case where the end item does have a large percentage of non-commodity purchased parts, the role of purchasing in the planning of such a new product becomes very important.

Second, the general approach and the extent of use of new policies, practices, and novel organizational structures will depend on the *significance of the project*. In the case of the Quad-4 engine, it was the case of a new engine design, informed significantly by motorcycle technology, and designed from the ground up. What is more, the processing plan called for self-qualification of each stage of the production and assembly process, which is described later in greater detail. This represented a radical shift in processing technology and practice for an automotive production plant.

Third, and finally, the case illustrates how an organizing theme—in this case a dedication to quality—can be a bridge between the development and implementation stages and a *continuing challenge to support incremental improvement* after the equipment vendors have left and production has accelerated. There was and is considerable variance in how these quality concepts are implemented by a variety of teams in the plant and the business units. The case shows how an entire quality function in an organization can be eliminated successfully by implementing effective process control as part of a new product launch program.

This report does not address in detail what was not done, which is as significant as what techniques were used to achieve the end result. For example, in product engineering, failure mode analysis and other formal methods of design analysis were intentionally not done. The time was spent on the design process, not analyzing the result. In supplier management, quotes were not obtained prior to source selection on selected parts. The time was spent evaluating the economics and capabilities of the suppliers. In manufacturing, receiving inspection was not planned from the start, which saved time and effort. Initially, there was no master list of control of critical part print characteristics. The time was spent on controlling the process.

But before the details of this program are discussed, the specification of this unique product are presented.

The Product Program

The Quad-4 engine has 2.3 liters of displacement using an in-line, 4 cylinder, 16 valve, iron block and aluminum head design. It produces 150 horsepower at 5,200 r/min. Compare these specifications to the 1955 Chevrolet V-8 engine introduced in Chapter One. Its maximum horsepower was 162 at 4400 r/min. The Quad-4 engine has 30 percent fewer parts than its nearest competitor, and in the process of organizing the program, the number of component suppliers was reduced from 200 to 69. The Quad-4 is pictured in Fig. 6.1.

The development team started with a clean sheet of paper for this project that had straight-forward, but challenging goals:

- Excellent fuel economy through low friction, good combustion, electronic fuel and ignition controls, and high-volumetric efficiency.
- High specific output to obtain desired power level at minimum en-

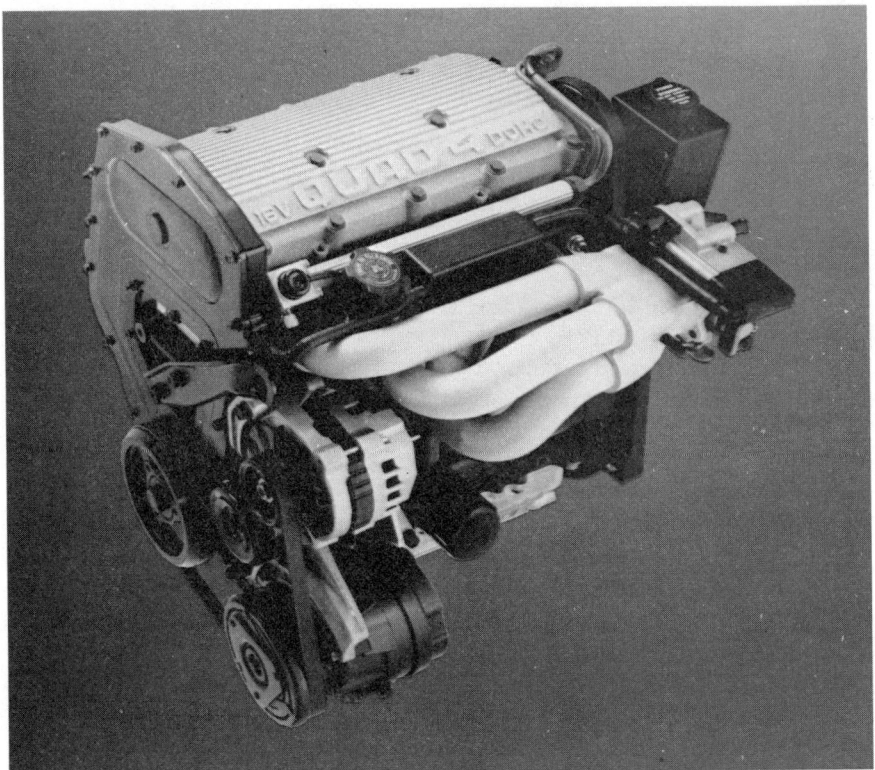

Figure 6.1 Quad-4 engine.

gine displacement with emphasis on low speed torque. This requires high thermal and mechanical efficiencies.

- World class quality, reliability, and durability achieved through: fundamental basic engine design, simplicity of engine control systems, capable, stable, and targeted manufacturing processes, teamwork among all disciplines, quality suppliers, and an extensive validation program.*

The engine concept was developed by examination of many types of existing technologies, including motorcycle engine products on the market at the time. The four-cylinder engine maximizes the technology available with the best value to the customer, it was thought, because it has the best fuel efficiency and power with the double-overhead cam. The displacement at 2.3L was optimum from the standpoint of packaging in a variety of GM transverse engine appli-

---

*From Thomson et al., 1987, p. 1.

cations. Additionally, the Quad-4 configuration is adaptable to emerging technologies. World-class durability and reliability was the goal.

In order to achieve these ambitious objectives, several basic philosophies became the guideline for implementation of the program. They included both constraints on the developing and implementing plant organizations as well as guidelines for purchased parts:

- Control of manufacturing variability at the point of manufacture. Reduction in part to part variation through:

    single sourcing
    target process characteristics at nominal tolerance

- Utilization of the team approach
- No 100 percent hot test facilities
- Simultaneous Engineering of design and process
- Continuous improvement in all aspects

Additional guidelines pertaining to purchased parts were:

- Early supplier involvement prior to final design (although basic design was known)
- Supplier capabilities assessed to determine potential of long term relationships
- Upfront supplier requirement identification and buy-in
- No in-plant receiving inspection
- Value (cost and quality) overriding factor
- Rationalization of the supplier base.*

In order to accomplish these guidelines, several novel approaches to organizing activity were taken. First, multi-disciplinary teams were formed with core membership having representation from product engineering, supplier management, manufacturing engineering, and quality control (which was a function later eliminated by the program after self-certification of quality was adopted). At one point or another during the process, representatives from service, packaging, reliability and test, finance, and product assurance were included. Perhaps the most unique feature of the Quad-4 project was the extensive supplier assessment and involvement program. The team was not obligated to work with any existing suppliers. This program and the spe-

---

*From Thomson, p. 3.

cial features of the manufacturing process are described in greater detail later.

A unique feature of the Quad-4 program was the product validation regimen. This program was designed to demonstrate and ensure world-class quality, reliability, and durability objectives. Focusing on various components and using reliability growth progress as the guide, an early precedent for continuous improvement was established in the program. In a way, this was the key link between design and manufacturing in this program. These are the guidelines for the Quad-4 engine validation program:

- Establish a validation program to include customer type usage and criteria as well as the traditional engine tests.
- Establish timetables to comprehend the pre-prototype, prototype, pre-pilot, and pilot phases of the program.
- Product content and specification changes monitored continuously and phased into the validation program in an orderly manner so that all design changes could be validated on time.
- Determine critical component reliability requirements and verify through supplemental bench testing.
- A "test incident" was defined as "any problem that a discerning customer would want to have repaired." An "incident" included such things as driveability complaints, stalls, oil leaks, as well as the traditional structural "concerns."
- Use a "reliability growth" measurement technique to monitor the progress of the validation program with respect to "world-class" goals to 100,000 miles.
- Set up an employee involvement feedback system so that all "Quad-4" engine concerns would be evaluated and resolved.
- Identify each incident, determine the root cause and validate the corrective action.*

To illustrate the extensiveness of the testing program, it should be noted that a total of 564 Quad-4 engines were built and tested during the program. A total of 307 vehicles accumulated 6.7 million miles, and 257 lab engines accumulated 50,000 hours of laboratory testing.

---

*From Thomson, p. 16.

Suppliers

Engine component suppliers

As mentioned previously, the number of component suppliers for the Quad-4 engine was reduced to 69 (compared to 200 for a typical engine of its type), because of the simplification of design, single sourcing, and adherence to fundamental engineering solutions, all of which substantially reduced the number of components. Because the engine was made up of mostly purchased components (approximately 70 percent), the success of the program was highly dependent on the quality of the supplier selection process. These companies were very carefully chosen by a multi-disciplined Oldsmobile team. At the onset of the Quad-4 project, the team was Oldsmobile but later it became part of the Buick-Oldsmobile-Cadillac Group or BOC, with a mission of expanding supplier roles to "partner-like" status. The old method of using multiple sources, based mostly on price considerations, was replaced by the idea of sole sourcing commodities, or large volume groups of similar components to those suppliers who offered top quality and timely delivery. Worldwide searches were performed for each commodity item, with each potential supplier being visited and methodically evaluated by the team. Allied and nonallied suppliers were both examined. While quality and timely delivery were the overriding criteria for selection, the desire and resources to work with Oldsmobile toward continuous improvement in quality and cost of components were a must.

The key point is that BOC wanted to monitor the assembly process for the engine, WITHOUT the need to worry about the dimensional accuracy of all the components. No supplied parts or products were to be inspected at the Delta Engine Plant, and no documentation regarding the parts would be accepted. Suppliers had to be regarded as partners in this 100 percent good practice (Huber, 1988).

Supplier responsibilities and selection process

This section summarizes the procedure that was used to evaluate and select the component suppliers for the Quad-4 engine, and describes the responsibilities that go along with that distinction. The starting point for the process was the BOC selection team.

The cross-functional teams from BOC, which were used to evaluate the suppliers for each "commodity," were made up of core representatives from purchasing, manufacturing engineering, product engineering, and what was then known as quality control. A simultaneous engineer had the responsibility for acting as the link between the design and manufacturing functions of BOC and the supplier, ensuring the compatibility of the component design. When needed, other expertise was brought into the team realm (product assurance, metallurgy, finance, and so forth). The whole process was very thorough and time-consuming. For example, nearly 100 foundries from around the world were evaluated as potential suppliers for the casting commodities. The evaluators got right down into the potential supplier's business, in order to make an informed decision. A chronological summary of the steps that were taken in selecting a component supplier and establishing a continuing relationship, as compiled from the references listed and supported by the interviews, follows.

Source assessment. The assessment process used by the team was a comprehensive survey of the supplier's facility focusing on the areas of quality, cost, management attitude and commitment, manufacturing capabilities and facilities, product development, people programs, and delivery. The survey was developed to assure consistency within each team and among teams. The selection team members spent the majority of their time on the manufacturing floor talking to the people involved in the product. Members of the team evaluated all aspects of the business from their own perspectives. To conclude an assessment period, the team discussed each item and reached a team consensus on ratings.

Assessment feedback to supplier. A key part of the process is the assessment feedback provided to all suppliers visited. Feedback was provided to the suppliers' management verbally (in a sit-down discussion) and in written form. This interaction supports a long term supplier relationship.

Identification of supplier requirements. Prior to finalizing the selection of Quad-4 suppliers, a booklet of supplier performance requirements and expectations was developed. The supplier manual of requirements/expectations included such items as sharing all cost data (pertinent to a part), implementation of a process control on all characteristics based on documented variability, packaging, and shipping guidelines. The primary emphasis was the delivery of quality parts. These were then discussed with the supplier in detail. If a supplier, for any reason, could not buy-in to all the requirements, the team approached the next supplier for that commodity. Buy-in to the philosophies and expectations was critical to success.

Control plan development. Each supplier was required to develop a control plan (with focus on continuous improvement) for each part that encompassed control of not only the print tolerances, but each process characteristic applicable to manufacture of the part. The control plan required identification of the measurement methods, documentation, and feedback systems for control of targetability and variability. Eventually the control plan moved from measuring the part characteristics to identification and control of the root causes of the process variables. The BOC personnel reviewed each control plan and recommended changes where appropriate.

Process validation. To ensure that the process was capable and the control plan was accurate, the suppliers were required to validate their processes. Short-term capability was documented prior to start of production through the GM sample submission procedure.

Continuous improvement. As mentioned earlier, a key concept in the program was continuous improvement. This applied to all aspects of the business whether design and performance of the engine, targetability of the processes, total cost to produce, including nonmanufacturing aspects of the business. Each supplier was expected to continuously reduce variation in the product, costs incurred, and improve performance to expectations through the life of the product. The use of Value Management techniques and training facilitated the improvement process, while institutionalizing a stable methodology.

There are several important points to note in the supplier selection process:

- A portion of a supplier evaluation survey form that was used has been reproduced in Appendix 6.1. Although originally targeted for suppliers of casted parts, it was quickly adapted for all suppliers into its present form.

- Suppliers are willing to share a lot of pricing and costing information usually considered proprietary because they don't have the hassle of having to get the contract every year. They, in fact, spread the costs over longer periods of time. GM realizes that suppliers have to make money to survive and are saying that's O.K.

- Documentation was used to spread knowledge within GM and transfer to follow-up projects (knowledge transfer).

- Morale improved. As one senior buyer put it, "The development of relationships with the supplier community does nothing but help. The understanding and commitment that are present on both sides really work together to make a quality product. You can take pride in it.... Before we would just buy the parts, and that was it. We're now intertwined and working together. We have a better understanding of the other guy's problems, and overall, you can't help but win."

- Communication channels now open that weren't before. For example, the critical role of purchasing (no longer sitting back, but active and involved).

- Need to evaluate suppliers within 24 hours of visit, and before the next one (discipline).

- Supplier becomes part of the Product Development Team... IMPROVEMENTS (MATERIALS, ETC.) ARE PURSUED TOGETHER. At the point of buy-in, the supplier officially becomes a member of the Product Development Team and assumes all the inherent responsibilities (Thompson, 1987).

- Elimination of receiving and inspection costs, which was crucial to the plan for eliminating the traditional quality control function.

- Problems:
  - *too many people swamping suppliers*
  - *initial search costs*
  - *geographical distance difficulties the best suppliers may pose (Italy is an example).*

Machine and technology suppliers

It was also critical for BOC to establish a close working relationship with all of their machine suppliers. The Quad-4 engine represented advances in the technologies of machining and assembly. The equipment came from various suppliers, such as Ingersoll, Cincinnati, Landis, IMPCO, Ex-Cell-O, and others. The fully automated engine assembly line was engineered and built by Comau Productivity Systems Inc. (Troy, MI) and its parent in Torino, Italy.

> The Comau fully automated engine assembly line was a first for the U.S. It is a palletized, nonsynchronous, power-rollered line, 2,056 ft long. It is made up of standard eight-ft modules, each independent with regard to operator and logic and power control.*

---

*From Huber, 1988, p. 49.

Role of product development teams

Multidisciplined BOC teams were again put together for the task of specifying, purchasing, and developing the equipment and the processes involved in building the engine.

> Teams were established for each process with engineers, supervisors, and hourly personnel all having input into the designing of all aspects of the process. Each team reviewed each operation in its process and determined level of control each one needed on each characteristic. The product engineer ensured that the specifications on the print were realistic for the manufacturing process and comprehended the requirements of the product.*

The manufacturing process engineers worked with the product engineers on the specifications. Hourly and salary work force input and involvement in machine and process design was started early in the equipment select, build, and runoff cycle. Multidiscipline teams were initiated early in the process planning stages and to maintain consistency, continued through machine runoff into production. Membership of these teams consisted of key employees from production, skilled trades, gauge and process engineering, and manufacturing and maintenance managers. These teams were initially formed and managed by the lead process engineer responsible for each part around which the team was structured. Hourly employees were mostly selected for the part and operation where they wished to participate. The teams became the cornerstone of the equipment runoff process, as well as a focal point in training and employee involvement in the Quad-4 program.

Buy-off procedures

The quality goals for the engine required that any machinery involved in its manufacture be capable of producing good parts. To assure this outcome, a lot of effort was put forth to define an appropriate, detailed buy-off procedure.

> Measurement systems were studied using a gauge repeatability and reproducibility study, and were required to meet established criteria. Process equipment was required to run parts in the equipment supplier's plant, with a specified capability and targeting after the process had proven stable.
> After the equipment was shipped and installed in the Delta Engine Plant, the measurement systems and equipment were verified again.‡

---

*From Blache et al., 1988, p. 26.
‡From Blache, p. 26.

A key element of the team building activity and project implementation was the specific machine "runoff" process utilized. In addition to accomplishing the required machine "runoff" goals, this enhanced team building with the machine tool supplier personnel has built a rapport which has continued on to machine installation and in-plant "runoff."*

Major suppliers still work closely with BOC personnel to continually upgrade the equipment and to improve its performance.

The Role of Quality in the Buyoff

Machine runoff was used to solve machine problems and to establish total process capability including machine and gauge capability. This was the first real team test, as well as an evaluation of the Statistical Process Control (SPC) system developed by these teams. Nowhere in the plant is 100 percent gauging used to sort "good" parts from "bad" (100 percent gauging is used only in conjunction with closed loop feedback control of tooling), nor are final inspection gauges used. The purpose of gauges is to assist in controlling the process variation This was part of the verification in process and product (VIP^2) system for quality and continuous improvement, which is taken up in depth in an upcoming section.

Plant Organization

The Delta Engine Plant uses shop floor teams to accomplish all production control (SPC), machining, assembly, and shipping tasks for the Quad-4 engine. In addition, the plant (with available floor space) is capable of expanding to produce other four-cylinder engines when it is time to produce additional products at the plant. In the current situation, the work flow allows for increased capacity introduction to double the production capability of the plant as planned acceleration and assembly plant requirements grow. The business unit concept extends to the maintenance and engineering function, as well (that is, equipment maintenance and process engineering are within the Production Business Units). The organization structure (see Fig. 6.2) has been developed to support three production business units (two machining and one assembly). The Manufacturing Services Unit (Central Maintenance, Industrial and Facilities Engineering, Tool Room, Tool Grind, Tool Design, Forward Planning, Electrical Engineering), Gauge Crib, Material Handling, Manufacturing Engineering, and Administrative Services support the Production Business Units.

---

*From Thomson, 1987, p. 15.

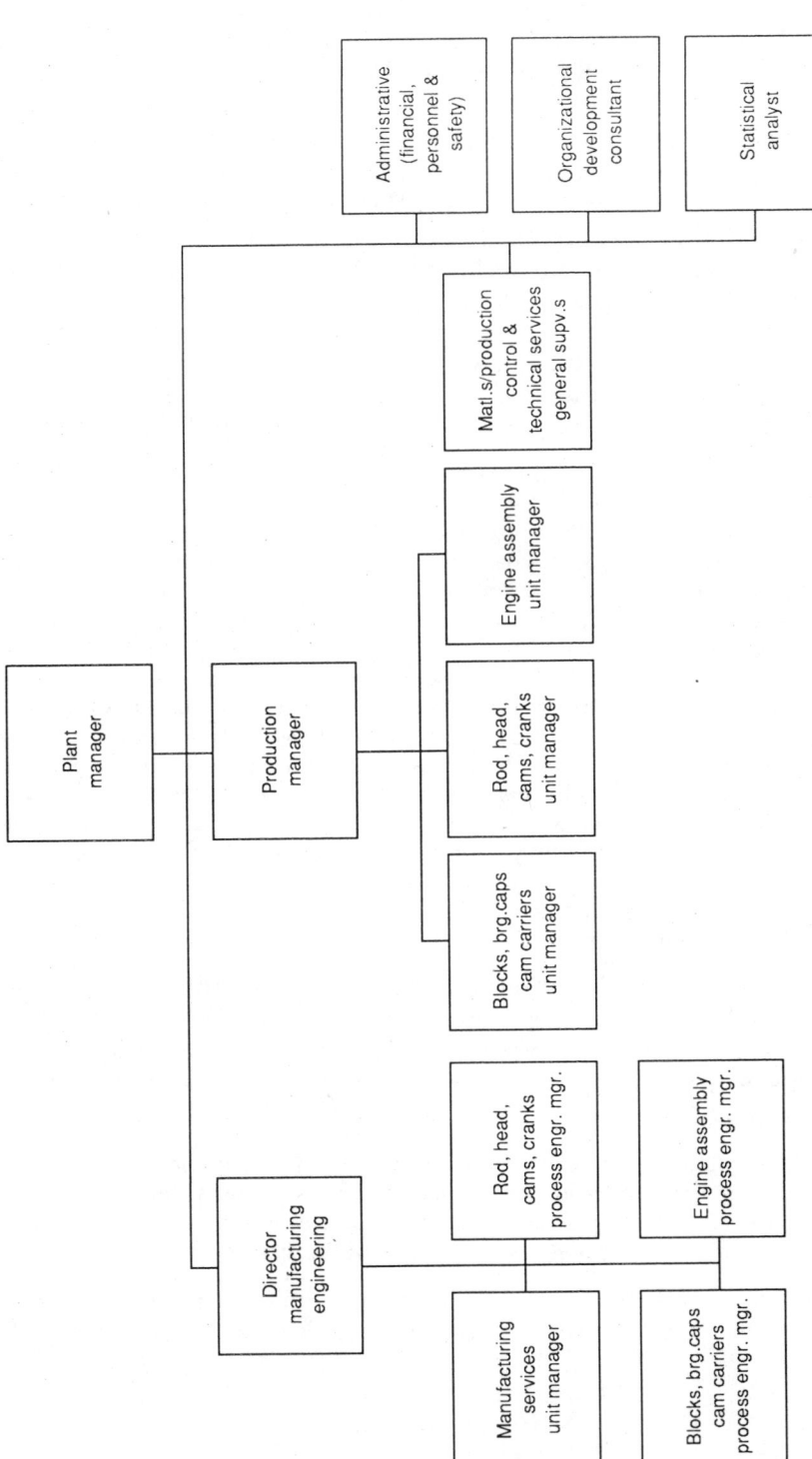

Figure 6.2 Delta plant organization.

The VIP² system for quality and continuous improvement

The Quad-4 project has changed the Delta Engine Plant operating philosophy from any previous precepts into one of "quality first." One very interesting and ironic result of the change toward this quality-first operating philosophy was the *elimination of the traditional quality control organization.* Instead of having an organization to monitor the production operators' work, a quality system under the acronym of VIP² (verification in process and product) was developed to monitor the engine's assembly. It supports control at the point of manufacturing and maintaining the shortest assembly feedback loop. The VIP² philosophy and procedure are detailed in this excerpt from IE.*

> It soon became obvious that the real need was to process the machining and assembly of the engine so that hot testing would not be required. (Note: "hot testing" is the process of starting the engine after it is built to make sure it is o.k.) The processing matrix that indicated on-line processes to verify that the engine was produced correctly at that point was called a VIP² station. These stations can best be described as just that: devices to measure variation...The VIP² system is not just a series of in-line gauges. It is more of a process or a method of assembly that has a series of physical stations. Its purpose is to ensure that the assembly process and product are correct at each point of assembly, as the product is being built. But to do this, VIP² has departed from the emphasis on part print specifications, and instead, focuses on assembly variation from a mean value—VIP² is designed to measure as many of those factors that are part of the total assembly process as possible. Some functional aspects of the supplier's work are measured, but the information is gathered mostly in addition to the main objective of measuring the assembly induced variation.

This new system allowed each operator to take on the responsibility for what was under his or her direct control, nothing more. An example of a VIP² station is illustrated in Fig. 6.3. Each operator now monitors his or her own operation, and takes appropriate steps to correct any problems he or she finds.

> The implementation of VIP² has been a study in patience and a test of maturity and dedication to an operation philosophy. It has been found useful to combine production, quality control, and process engineering disciplines in implementing the details of the VIP² system.‡

The successful transition to this new quality philosophy was strongly dependent on the work of many multidisciplined teams such

---

*From Blache, p. 27.
‡From Blache, p. 28.

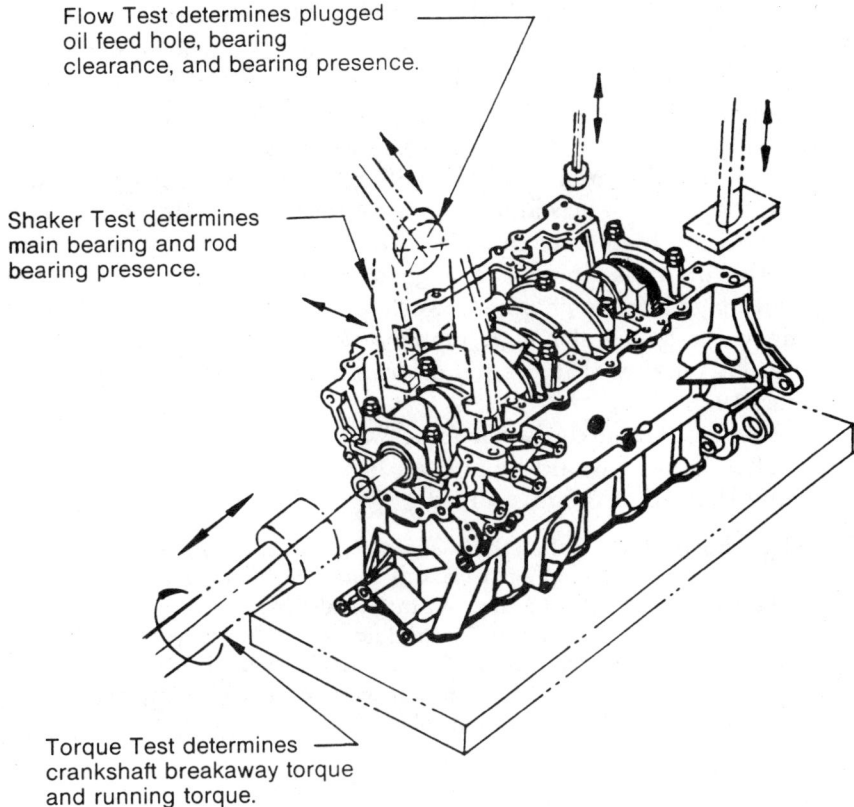

Figure 6.3 Verification in product and process (VIP^2) station.

as those that were formed within the Delta Engine Plant. Examples are cited throughout this case. The point is that a large, concerted, unified effort was necessary to ensure the success of this project.

Team building

Various examples of the team oriented approach that was used throughout this project are described in this case. However, there are some common themes that pervade the whole situation of how the teams operate and the environment in which they do so.

First of all, the management team at the Delta Engine Plant has taken the attitude that the technology must be implemented in such a way as to keep the employees the "dominant factor in the production process" (Blache, 1988). Despite the rush to operatorless factories in the manufacturing world, 77 of the 160 work stations on the Quad-4 assembly line are manual (Huber, 1988). Many of those involved report that the level of job satisfaction they feel is much higher because of this project and the increased responsibility it has brought.

Team oriented activities have been successful at the Delta Engine plant because of many things, not the least of which is the caliber of personnel at the plant. Many different personalities have been brought together to accomplish various tasks. An additional plus has been the good history of labor relations that the plant has enjoyed. Since the plant is known as a good place to work, a lot of high-seniority people have chosen to work there. This has helped stability in the work force, although employees can move between several plants. This makes things such as training a lot easier to implement (Huber, 1988).

However talented the work force was at the outset of the project, their effectiveness would have been lessened if not for the extensive training that all of them received.

> Training in statistical methods and statistical thinking were particularly important for developing a critical mass for continuous improvement. Classes such as Introduction to Statistical Process Control, The University of Tennessee's Institute for Productivity Through Quality, Basic Statistics, ANOVA, Design of Experiment, Taguchi Methods, and W. Edwards Deming's four-day seminar were attended by employees as they needed the training to do their jobs.*

Additional observations on teams

The underlying factor to all the work that was accomplished with teamwork in this project is the communication process. Most of those interviewed saw effective communication as the major ingredient to a successful team effort. A chief engineer emphasized the need for all groups to be given adequate information to make a decision. Withholding information from any team, unintentional or otherwise, can produce disconcerting situations. There was an instance of a group's original decision, made by consensus, having to be overturned by a superior because a meaningful bit of information important to the decision was not available to the team. Once empowered with the authority to make a decision, it is critical that the team be supplied with any and all information pertaining to that decision. This will avoid uncomfortable situations that may damage credibility and the team's effectiveness.

As successful as the teamwork concepts and practices have been, the Delta Engine Plant is not without its challenges. Themes such as "fo-

---

*From Blache, 1988, p. 26.

cus on the operator," "no non-value added work," "do it right the first time," and "decision making at the lowest level," take management consistency, time, and self-control to become the accepted plant culture. The charter quality plan for the plant calls for continuous improvement in all aspects of the business. Sustaining the success enjoyed so far will require a large effort on behalf of all those involved.

Outcomes of the Quad-4 Program

This case history benefits from the fact that actual performance and market impact outcomes are available. Given the climate of today's automotive markets and competition, perhaps the most important indicators of success or failure of a project are the quality and market acceptance or sales figures. But overall time to launch the product is also important, if new introductions are going to respond to markets. Therefore, first we compare the Quad-4 with three other comparable engines launched at similar times (see Fig. 6.4). Two of the competitor engines are Japanese and one is West German. The launch times from

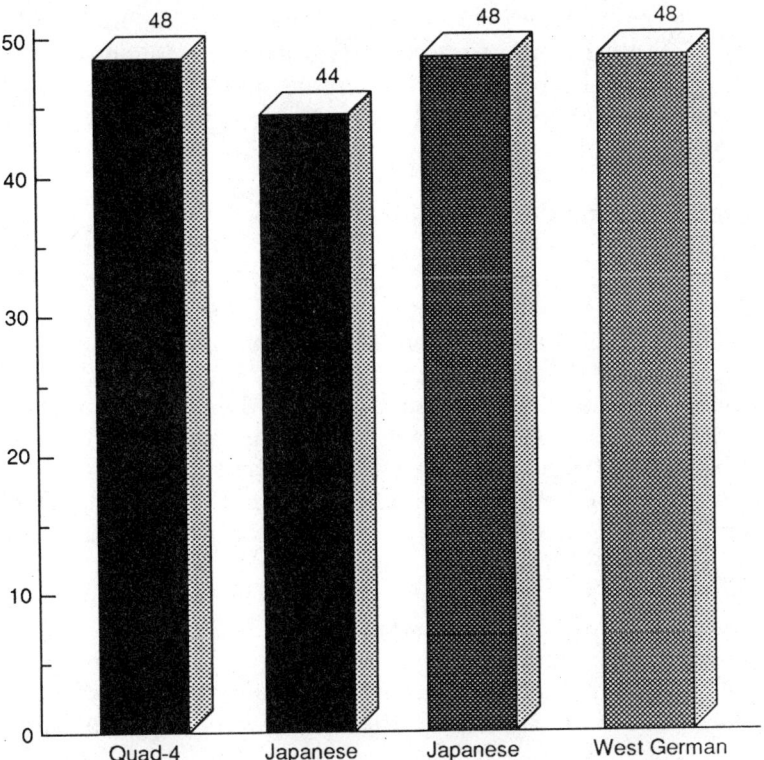

Figure 6.4 Concept-to-car launch time of four engines.

concept to production engines are nearly identical. The Quad-4 required a total of 48 months to launch—33 months from approval to production engines and 15 months to develop a successful approval. Note that one of the comparable Japanese engines required 44 months to launch, and the other Japanese engine, as well as the West German engine, required an actual total elapsed time of 48 months.

We review two types of market information here. First, the results of the customer survey data or CAMIP—Continuous Automotive Market Information Program—as it is called at GM which is the customer approval rating used internally (see Table 6.1). Second, we report internal quality results.

The CAMIP data on a total of eight engines—the Quad-4, five other domestic engines, and two Japanese engines—are presented in Table 6.1. For both the first quarter 1988 (at 99 percent beating all competitors) and third quarter 1988 (at 95 percent tied with one Japanese competitor), the Quad-4 has the highest overall rating. The Quad-4 also had the fewest problems per 100 engines in the third quarter, and was bested only by one Japanese engine in the first quarter (26 versus others, and 21 versus 13, respectively). These results were all achieved during the first year of the Quad-4 launch. For 1989, the first quarter showed the Quad-4 in first place (tied with one Japanese competitor) at 96 percent. Problems per 100 engines were at 16, compared to the best Japanese competitor at 11. The sustained positive customer response is evidence of a stable process and "doing it right the first time."

In addition, internal GM estimates show that the Quad-4 engine contributed an additional 8000 plus sales units to GM through April 1988, which otherwise would not have been sold. In other words, the engine is having a significant impact on purchase decisions. About 19 percent of Calais Quad-4 buyers name an import model as their second choice as opposed to only 9 percent of Calais buyers without the Quad-4 option. Over 50 percent of the Cutlass Calais models are sold with the Quad-4 option. It agrees with the remarkable 99 percent CAMIP rating for the first quarter 1988 earned by the Quad-4.

Plant data on internal audits for quality confirm these trends. During the last three plant audits, which are generally done monthly, through the period ending in 1988 calendar year, the plant has had three perfect quality audits for controllable problems. Base engine (Quad-4 plant build responsibility) warranty costs are down, and as mentioned earlier, even without a hot test in the Delta Engine Plant, the number of engine pulls in final assembly—which were very low at the outset—have actually decreased. Keep in mind that this is being attained with typical process clearances (actual performance) of 0.001 mm to 0.007 mm piston pin to piston pin bore, 0.035 mm to 0.038 mm

TABLE 6.1 Continuous Automotive Market Information Program (CAMIP) Data on Eight Engines

Parameter	Quad-4 L4-2.3L	Domestic L4-2.5L	Domestic L4-2.0L	Domestic L4-2.0L	Domestic L4-2.3L	Domestic L4-2.5L	Japanese L4-2.0L	Japanese L4-2.0L
			1988 1st CAMIP—L4 Engines					
Overall Engine	99	89	89	87	80	90	97	97
Prob/100 Eng.	21	38	45	51	55	32	13	31
			1988 3rd Quarter CAMIP—L4 Engines					
Parameter	Quad-4 L4-2.3L	Domestic L4-2.5L	Domestic L4-2.0L	Domestic L4-2.0L	Domestic L4-2.3L	Domestic L4-2.5L	Japanese L4-2.0L	Japanese L4-2.0L
Overall Engine	95	90	84	88	86	90	94	95
Prob/100 Eng.	26	59	65	47	71	41	29	27

SOURCE: Reprinted, by permission, from General Motors Corporation.

valve lifter to lifter bore, and 0.015 (6 sigma) connecting rod crank end diameter.

Overall, this product has had an impressive launch. In terms of both internal and external measures, the Quad-4 has been an early success for GM. It's an excellent example of "do it right and don't look back." One respondent put it rather succinctly: "After the Quad-4, you could go to a social gathering and be proud to be a GM employee." The numbers clearly support this conclusion.

By the same token, the project represented a significant investment by GM. No engine has ever been tested like the Quad-4. Engines were tested to the equivalent of running a car 100 miles per hour for 100,000 miles. Recapture of these high costs represents a significant corporate technology transfer challenge for GM. But efforts are underway so that this will continue—and not by chance. For example, the results of the Quad-4 engine testing program will allow other new engine programs at GM to test only for critical performance parameters. This represents a significant cost saving with no loss in quality and performance. The business units, plant, group, and total organization have been briefed on the outcomes of using groups and multi-disciplinary teams to launch a major new product. This sharing of learning experience can provide some of the returns on this project that would otherwise not be available.

Besides the formidable corporate technology transfer challenge that the Quad-4 engine project presents to GM, there are two other continuing challenges on this project. For one, the replication of a Quad-4 type experience elsewhere in the firm is a test of the company. That is, can it be done across other divisions and for other products that have not had a tradition of close product-engineering coordination? For another, the sustaining of the Quad-4 effort in the Delta Engine Plant is also a real challenge. The product demand cycle goes up and down, and it puts a real strain on the program. Sustaining the team concept with personnel turnover in a shop floor business unit structure is not a unique problem to this facility, but it is real. Allocation of resources across these business units is continually reappraised. But constant here is the Quad-4 product and its identity in the Delta Engine Plant. This benefit is difficult to quantify but contributes to the prospects for successful plant management, nonetheless.

Appendix 6.1 BOC-Powertrain Delta Engine Supplier Survey

COMPANY_____
LOCATION_____
DATE OF VISIT_____

	RATING AVERAGE	WEIGHT	WEIGHTED RATING
A. GENERAL INFORMATION	_____	_____	_____
B. GENERAL QUALITY INFORMATION	_____	_____	_____
C. PRODUCT DEVELOPMENT	_____	_____	_____
D. RECEIVING AND SHIPPING	_____	_____	_____
E. MANUFACTURING SUPPORT	_____	_____	_____
F. LABORATORY SUPPORT	_____	_____	_____
G. MANUFACTURING FACILITIES	_____	_____	_____
H. SECONDARY OPERATIONS	_____	_____	_____
I. PRODUCTION PART CAPABILITY	_____	_____	_____
J. FLOOR INSPECTION EQUIPMENT	_____	_____	_____
K. PROCESS CONTROL	_____	_____	_____
L. CONTINUAL IMPROVEMENT AND RECORDS	_____	_____	_____
M. ATTITUDES	_____	_____	_____
N. TRAINING	_____	_____	_____
O. OVERALL GUT FEEL	_____	_____	_____

RATED AVERAGE _____

TEAM MEMBERS: Purchasing_____
                   Quality Control_____
                   Product Engineering_____
                   Manufacturing Engineering_____

A. GENERAL INFORMATION	RATING	WEIGHT	WEIGHTED RATING

1. Management Involvement
   a. Management aware of quality measurements scrap rates, etc.
   b. Unified and consistent policies
	_____	_____	_____

2. Expansion Plans
   a. Facilities
   b. Support equipment
   c. Production equipment
   d. Personnel —management
              —labor

3. Sales Representatives
   a. Availability
   b. Past responsiveness
   c. Technically capable
   d. Informed and knowledgeable

4. Labor Contract
   a. Union
   b. Length of contract
   c. Expiration date
   d. Past strikes

5. Employe Involvement
   a. Knows productivity and quality status
   b. Informed on business matters
   c. Participates in setting goals

6. Employe Incentives
   a. Profit sharing
   b. Incentive to remove and report defects
   c. Employe recognition and service awards

7. Employe Support
   a. QWL programs—formal and informal
   b. Day-care centers, flex time
   c. Drug/alcohol programs
   d. Organized outside activities
   e. Company supported outside facilities
   f. Educational support

RATING AVERAGE _____

B. GENERAL QUALITY INFORMATION	RATING	WEIGHT	WEIGHTED RATING
1. Spear a. Rating_____ b. When awarded_____ c. Last rating_____ d. From what GM organization_____			

2. Programs to improve ratings_____

3. SQI
   a. Rating_____

   b. Reasons:_____

4. Internal scrap rates
   From documentation, not verbal
     information.
   OVERALL
   PART a. _____ VOLUME ____ RATE
   PART b. _____ VOLUME ____ RATE
   PART c. _____ VOLUME ____ RATE

5. Customer return rates
   Consider complexity of parts
   OVERALL
   Part
     a. _____ VOLUME ____ RATE
   Receiving inspection at customer
   PART
     b. _____ VOLUME ____ RATE
   Receiving inspection at customer

                           RATING AVERAGE _____

	RATING	WEIGHTED	WEIGHTED RATING
C. PRODUCT DEVELOPMENT			
1. Product Design Personnel			
a. Experience of technical organization			
b. Input well written and documented	_____	_____	_____
2. Product Design Facilities			
a. Design from functional requirements			
b. CAD/analyze and refine the design			
c. Finite element analysis			
d. Modeling			
e. Testing facilities			
f. Utilization of facilities	_____	_____	_____
3. Prototype Support			
a. Do you have prototype facilities			
b. Short lead times and prompt deliveries			
c. Prototype parts off production tooling			
d. Prototyping normal business practice	_____	_____	_____
4. Research and Development			
a. Design innovation projects			
b. Areas of research and development			
c. Patents and proprietary designs	_____	_____	_____

5. Estimating
   a. Process well documented
   b. Information clear and correct
   c. What is the average time to quote

RATING AVERAGE _____

E. MANUFACTURING SUPPORT	RATING	WEIGHTED	WEIGHTED RATING

1. Manufacturing Engineering Support
  a. Process engineering
  b. Gauge engineering
  c. Tool designers
  d. Plant engineers			

2. Skilled Trades
  a. Electrical, mechanical, tool & die
  b. Formal (apprentice) training program
  c. Continually updating skills
  d. Evidence of professional repair practices			

3. Machine Support Facilities
  a. Machine shop—neat, clean, well equipped
  b. Crib—machine parts
  c. Refurbishing/rebuild area
  d. Prints/manuals
  e. Electrical Drawings			

4. Tooling/Die Support—Central
  a. Tool inspection
  b. Tool crib
  c. Tool sharpening facilities
  d. Tool records
    • Tool life
    • Problems/performance
    • Prints			

5. Preventative Maintenance System
  a. Lubrication
  b. Fixture
  c. Tool change
  d. Records available and utilized
     Down time
     Machine problems			

RATING AVERAGE _____

References

Blache, Klaus M., Stewart, Kenneth C., Zimmerman, Robert L., Shaull, James E., Benner, Raymon D., and Humphrey, Gary P., "Process Control and People at General Motors' Delta Engineer Plant," *Industrial Engineering,* March 1988, pp. 24–30.

Huber, Robert F., "VIP = Very Impressive Plant," *Production,* November 1988, pp. 48–52.

Miles, Donald L., and Waldron, William A., "The Quad-4 Engine Program," presented in the Workshop, Managing New Technology in Manufacturing, The Eight International Automotive Industry Conference, The University of Michigan, Ann Arbor, Mich., March 23, 1988.

Thomson, M. W., Frelund, A. R., Pallas, M. and Miller, K. D., "General Motors 2.3L Quad 4 Engine," SAE Technical Paper 87035, The Society of Automotive Engineers, International Congress and Exposition, Detroit, Mich., February 23–27, 1987, p. 19.

Chapter

7

Approaches to Product-Process Development Management

James N. Hughes*

Introduction

The design of an organization for conducting concurrent product and process development demands at least as much attention, skill, and effort as the design of the products themselves. Meeting current intense competitive pressures on time and performance with inherited work practices and hierarchies unchanged for decades has become increasingly difficult. Organizational barriers to fast effective product-process design† can and should be removed.

The theme of this chapter is to define a path to improved management of product-process development. This path represents a transition from the "old way" to a "new way" (see Fig. 7.1). Making this transition is no easy task, but the improvements in product-process development as well as overall organizational effectiveness are well worth the effort. In this chapter, we present a vision of the new way and discuss possible approaches for achieving it. We do this by drawing selectively on certain related project experience within the variety of techniques applied by GE and other companies. According to *Business Week,* since 1981, GE has used integrated design in more than 100 development programs, from major appliances to gearboxes for jet engines. GE estimates that the concept has netted millions of dollars

---

*This chapter is based on the author's 37 years of experience at GE and other firms. The opinions here are those of the author and do not necessarily reflect the official policies of GE.

†Throughout this chapter, "product-process design" will be referred to as a single entity, to emphasize integration.

THE OLD WAY	THE NEW WAY
1. Organizational hierarchy of compartmentalized functions based on educational specialization: Engineering, Manufacturing, Marketing	1. Organization based on work flow, with minimal overhead layers, and groupings of work to reduce time cycles.
2. Barriers to communication and integration between functions.	2. Needed information channels are built into the system, as closely coupled as possible.
3. Separation from customers and market/competitor actions, coupled with internal measurements.	3. Customer/market focus is built into the sequence of work and the measurements.
4. Many approvals, many transactions, many interfaces.	4. Fewer interfaces, easier definition, and more operating autonomy.
5. Preservation of organization is dominant. Introducing change is difficult.	5. Change is dominant, the organization is more flexible and agile.

Figure 7.1 The "Old Way" versus the "New Way."

in benefits, either from cost savings or increased market shares (*Business Week,* 1989). Many of the applications discussed are site-and-situation specific, but, in all cases, the common thrust is to define ways to match people's activity more closely to both customer needs and to a high rate of technical change. Traveling the path to improved management of product-process development is obviously a journey, not an event, but whoever nears the end of the trip first will have major advantages over those who have just begun.

Organizing for Change

Traditional ways of grouping people's efforts in an established organizational structure have operated well in a relatively stable environment. Emergence of global competitors has altered this environment to one of continual change, trending always toward more product value being delivered with less use of resources. Traditional functional organizations have shown difficulty in adapting to this high rate of change. The challenge is to achieve reasonable core business stability and growth in the face of increasing variability in the demands made by customers or forced by competitors.

Product-process change is fueled by the customer's ever increasing appetite for variety. Consumer products are driven by individualism and a desire for products that are "personalized" as opposed to "mass-produced." Industrial product change is driven by the user's eternal need for more effectiveness, by new technology, and by increasing desire for custom-designed solutions.

Now add the emerging critical factor of *time*. With increasing competition on a global basis, the products and features first into the market have higher profit potential. Whoever is second can expect less volume and most likely lower margins. The combination of these factors—variety and time—makes it essential that companies define new products and processes more frequently, quicker, and more perfectly than ever before, from all viewpoints both technical and economic. This dictates that the organization must focus on product-process change as the dominant driver.*

Two classes of change

Integrating product-process changes into the organization involves two classes of change: (1) product-process change itself, and (2) shaping and changing the organization to accomplish product-process change more effectively. Dealing successfully with the first class of change requires at least the following:

1. Deciding which product features, at what timing of introduction into the market, and at what feasible cost margin demands the most competent minds in the business, working together in balance and with a common plan. Determining "WHAT" is not a matter of delegation to lower levels.
2. To move quickly, relevant research, advanced technology, inventions, and other unpredictable activity needs to be done *before* the product-process design, so that product design itself is a more controllable process.
3. Process design (how to make it) must be totally interrelated with product design, from concept onward, to improve the product, to continually improve costs, and to improve value to the customer.

The second class of change, that of shaping and changing the organization itself to accomplish the first class of change more effectively,

---

*"New Product Introduction" may be defined to include:
1. Analyzing the market and defining specific detailed customer and product strategy to preempt or neutralize competitors
2. Initiating the design process with an appropriate specification to meet *both* business and product targets
3. Developing integrated detailed technical definition of products and appropriate processes to meet the specifications
4. Providing facilities and information to start-up and produce the new products
5. Producing, delivering, and reaching customer satisfaction with the product.

should be the subject of management action. In considering how the organization should be shaped to enable and facilitate product-process change, we offer the following insights:

1. The necessary work activities for new product-process changes should be the backbone of the organization. All else is support and can frequently be done with lesser urgency, or at least in parallel, and at less cost, by relocating and consolidating.
2. Grouping these work activities in their most effective sequence and flow provides a direct organization structure for introducing change.
3. Grouping by work flow leads to more autonomous work groups, clearer operational targets and measurements, clearer interfaces, and shorter cycles, whether in development or in production.

Achieving a change oriented organization

The purpose of structuring the work effort around change introduction is to maximize the focus on product-process technology, which exists principally to continually prepare the business to supply appropriate products. How can an existing organization analyze its current activities and develop its own unique approach to a more change-oriented, flexible, low-overhead, quick-reflex business? One approach being used at firms, called Work Flow Analysis, is centered on diagramming or charting the path taken by new product programs during development and introduction and quantifying the activity effort being applied.

Work Flow Analysis involves determining and plotting in sequence the existing steps in a particular process (see Fig. 7.2). Insight comes from using a technique of viewing each action step in terms of "input/action/output" such that the output of a given action constitutes the input to the next action. Bringing into view the numerous loops, sequences, approval chains, interlocked transactions, and rework items that exist in most businesses can be unsettling. However, recognizing the complex actual path of work is a necessary first step in dealing with it.

Having arrived at some data on how time is currently being spent, a next step is to make a concerted effort to reduce the detail times. A helpful method is to visualize the usefulness of activities in contribut-

---

Figure 7.2 Steps toward organizing work flow: (*a*) define activities that cover an organization's work, and quantify collected data to show where they are performed; (*b*) prepare a work flow chart showing current operations and grouping work into input/output sequence; (*c*) sort activity data into major work groups identified, quantifying the work application

```
: CODE :            ACTIVITY/DEFINITION
...................................................
:   14 : BUDGETS AND FORECASTS
:      : Prepare program budgets and forecasts.
:      :
:  122 : BUILD TOOLS, TEST EQUIPMENT, JIGS, ETC.
:      : Design, build, assemble, and modify equipment, jigs,
:      : fixtures, tools, etc., used in    assembly and
:      : test.
:      :
:    8 : COMMUNICATIONS - STATUS AND PRIORITIES
:      : Communicate program schedules, status, priorities
:      : and decisions to own and other supporting
:      : employees.  Includes preparation of presentations
:      : and writing regular activity reports, IOI's, etc.
:      :
:   59 : COMPONENT DEVELOPMENT INSTALLATION & TESTING
:      : Assemble, install, instrument, and checkout
:      : component test rig.  Perform testing, obtain and
:      : evaluate results, change test plan as required to
:      : attain final test objectives.  Present data.
:      :
:   62 : COMPUTER AIDED ENGINEERING INTEGRATION
:      : Coordinate with CAE to plan and develop new software
:      : and hardware for data processing.
:      :
:   19 : COMPUTER AIDED TESTING SYSTEMS ACQUISITION
:      : Acquire and evaluate new or revised applications for
:      : automated testing.  Includes equipment and software
:      : evaluation.
:      :
:   20 : COMPUTER AIDED TESTING SYSTEMS DESIGN
:      : Specify, analyze and design new or revised
:      : applications for automated testing.  Includes
:      : equipment and software evaluation for CAT.
```

(a)

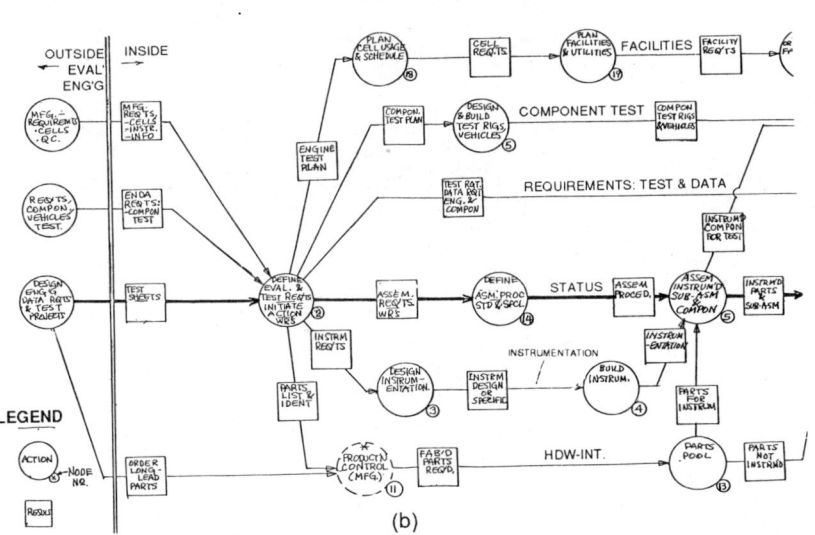

(b)

MODE 2 — DEFINE EVAL. & TEST REQTS, & INITIATE ACTION

ACT NUM	ACTIVITY NAME	TOT ACT PEOPLE	TOT EQUIV PEOPLE	3-4 EQUIV PEOPLE	3-6 EQUIV PEOPLE	7-4 EQUIV PEOPLE	7-6 EQUIV PEOPLE	7-8 EQUIV PEOPLE	8-8 EQUIV PEOPLE	9-4 EQUIV PEOPLE
	MODE-SPECIFIC ACTS									
5	MANAGE PROGRAMS	10	1.5	0.0		0.1	0.2	0.2	0.2	0.1
54	ASSEMBLY AREA SUPPORT		0.4			0.0	0.1	0.1	0.0	0.0
57	ASSEMBLY PROCEDURES	30	3.3			0.8	0.4	0.5		0.3
63	COMP CONSOLE PROGRAMMING	2	1.1							
64	DATA ACQ HARDWARE DEV	12	4.1	3.8	0.3					
80	TEST-SETUP ENDURANCE	18	1.8			0.3	0.1		0.3	
81	TEST-SETUP NON-ENDURNC	18	1.5			0.3	0.2	0.0	0.3	0.4
83	ENGINEERING DOCUMENTATION	50	4.1		2.1	0.0	0.2	0.0	0.3	0.4
86	ENGNG TEST - PLANNING	26	1.8			0.3	0.1	0.3		0.3
90	ENGNG TEST WORK REQUEST	25	0.1			0.1	0.0		0.1	0.3
100	POST-TST HDWR DSP-CUST REV	15	0.8			0.2	0.1	0.3		0.4
101	POST-TEST HARDWARE INSPECTION	29	1.5				0.3		0.3	0.8
112	TEST RESULTS - REPORTING	33	2.6						0.1	0.2
164	FLD FAILURE -INVEST	17	1.8							
			33.5	3.8	2.4	3.6	2.7	2.1	1.7	4.4

(c)

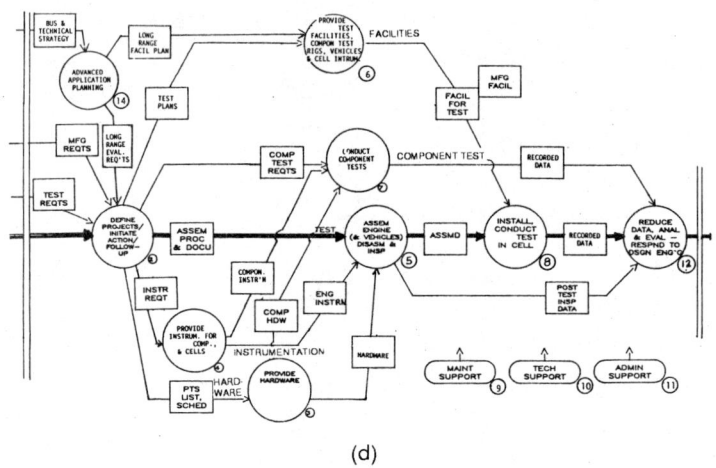

(d)

EVALUATION ENGINEERING

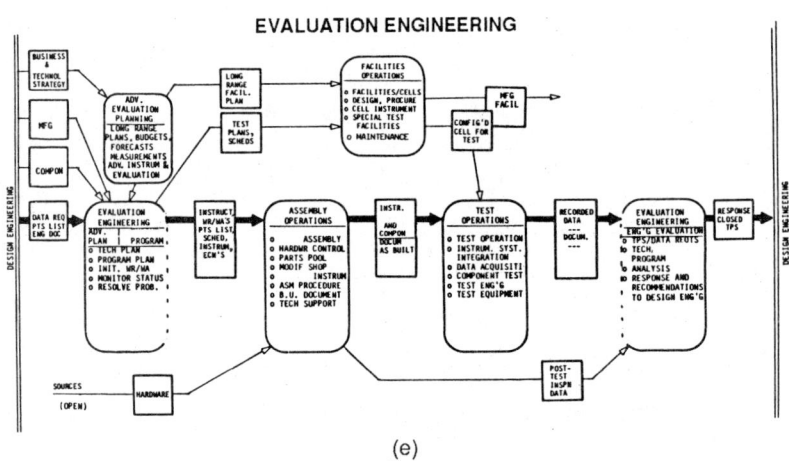

(e)

Work Group Guide
Overview

Work Group: Test Operations

Action:
Operate test facilities, including installing components and assemblies for test, installing instrumentation, collecting prescribed data, conducting specific test cycles/sequences, and logging data events needed for evaluation.

Input:
Test articles provided on schedule with special instrumentation, non-cell instrumentation, test procedure, data logging requirements, schedule requirements, specific evaluation items required. Administrative and budget targets and decision elements.

Output:
Completed tests on components and assemblies documented for evaluation, on schedule, in format required, per test procedure. Observations and recorded data to provide engineering with maximum meaningful insight into performance as defined in test instructions.

(f)

ing value to the final product. Those activities that stand out as having low value should be questioned for elimination or combination. Certainly, time elapsed with no contribution does not add value. Another approach is that any activity that one can perform in an alternate or better way at less cost or time is not adding enough value and should be replaced by a better alternative.

Using the Work Flow Analysis, an organization alignment and grouping is created where the work output of each work group is organized to be fully accomplished as the input to the next work group. A majority of support work is moved off-line or in parallel so that only necessary actions are performed on the main line resulting in a shortened total time cycle. Grouping the work activities in this defined manner supports continual process improvement and measurement of the output. Process perfection on the main line contributes to better quality and "doing it right the first time" whether in design and development, in production, or in office administrative processes.

The support group activities are needed in varying degrees to keep the main line operating, but add value to the product only indirectly (for example, utilities and equipment maintenance) or not at all (for example, janitorial or security force). The challenge is to reduce these costs to a minimum.

The end point being sought is to achieve a practical sequence of work groups that have the shortest overall time to completion but without shortchanging any specific activity. In fact, the resources needed for improving quality may well be generated by eliminating the unneeded or low-value activities or by finding alternative methods.

A key part of the main line work sequence analysis is to rearrange specific tasks to eliminate dead time spent waiting for other tasks to be completed. This has conceptual elements of critical path charting but is usually less rigorously applied because of the complexity of issues. For reference, the average actual time normally taken between current steps in the process should be documented to compile a time-network of the major steps.

Since paralleling of activities is desirable to save overall elapsed time, the question exists of matching the results when they come together, such as at the interfaces. Studious attention to interface definition early in a development program can help. One illustrative GE experience was in developing complex cabinets of electronics that were later interconnected. Too often the final system assembly turned up incorrect connec-

---

Figure 7.2 (*Continued*) (*d*) analyze work flow to develop a direct Main Line of needed work and expose duplication and low-value work; (*e*) refine work groupings to establish self-sufficient Main Line work groups, interfaces, and off-line support; (*f*) define and detail work group responsibilities, staffing, and training to implement.

tions, mismatched cables, and introduced other irritants at the most inopportune time. A helpful tactic was to define cable connections first, thus firming up the interfaces at the outset. Although this took time and discipline initially, it saved time in the long run (see Chap. 4).

An organization based on work flow

Let's consider what kind of organization would be the most successful at product-process change. Of necessity, the organization would be structured to provide maximum variety at minimal overhead cost. The central focus would be to provide strong integration of change-introduction, centralized tactical command, and rapid delivery of orders. Formal structure would be held to a minimum to maximize flexibility and quick response. Consequently, we would expect only a few large groupings of activities centered on the work flow. *Every* element of work must add value to the product—either to the one being shipped today or to the next one being designed.

What might the organization chart for such a "lean and mean" organization look like? One answer to this question can be found by identifying the most basic *actions* that must take place to conduct the business. Four groupings of "main line" work appear logical: decide what customers want, set up the factory to make it, produce it, and sell and service it. All other functions can be moved off the main line and consolidated into a support grouping. How would these work groupings be structured in the simplest work flow possible? Tracing the generic flow of work within a business yields two major paths: (1) receipt of orders and production of existing products, and (2) planning and introducing new product changes into production. Placing the basic work groups in the context of this work flow provides the organization chart we are looking for (see Fig. 7.3).

Such an organization alignment maximizes the dynamics of change, while working to minimize the static overhead and shorten both the order to delivery cycle and the change-introduction cycle. Note that the organization chart (see Fig. 7.3) is deliberately *not* stated as blocks and lines, which imply function command from above and function compliance below. The interfaces are working interchanges dedicated to providing products to customers. Although site-specific, each subelement would be structured along the following lines.

Management tactical planning. The management tactical planning group is the core business management and staff—the most experienced people—who direct the course of the business. They determine the "what" and "when" which, on balance, are the best product actions the business can make, and quantify the needs that guide the techni-

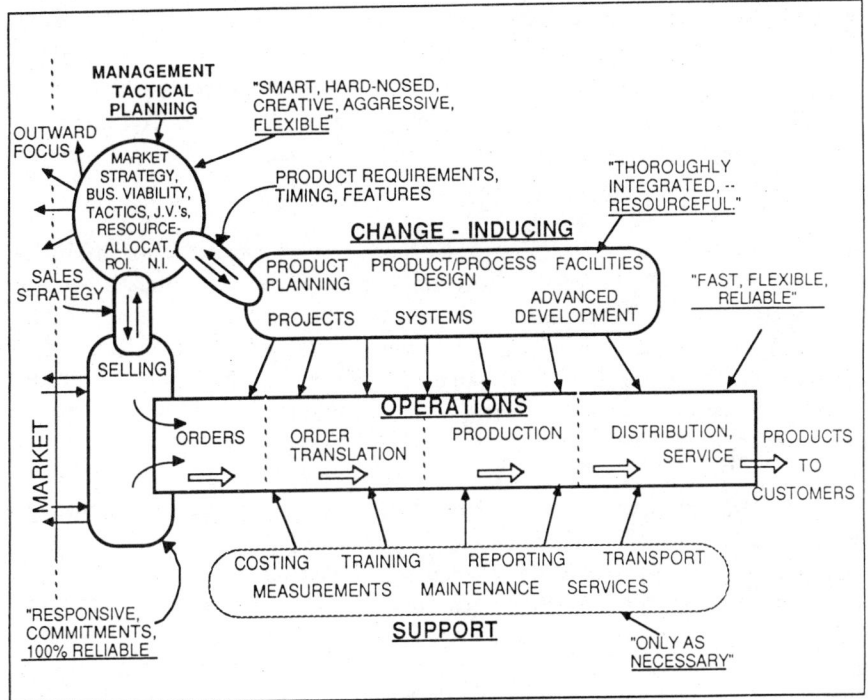

Figure 7.3 Organizing for change.

cal areas. Operation of the group is more "working" than "presiding," there is less remoteness from middle managers, and there is on-going swapping of roles and actions to stay current. Group characteristics and functionalities include the following:

- Both business and technical expertise is included, tightly intermixed, with cumulative judgment to sense customer trends, evaluate competitors, assess resources, and define what to do. The emphasis is not on preserving status, but on deciding effective action by working diligently.
- Work practices are more like those of middle managers, and less like boards of directors. The tenor is rapid and the team interplay is vital.
- Closely associated data-gathering, decision-support systems, comprehensive data bases, and information resources are a given.
- The output of the tactical planning group consists of (1) product requirements (that is, detailed specs for features, performance, compatibility, ergonomics, and appearance or styling), and (2) cost and margin, unit volume, expected capacity utilization, and timing of in-

troduction. Nominally, these items have been called "product specification" and "business plan," but the intention here is to achieve more muscular programs and more integrated attention to being first into the market with the right product. Additionally, close attention is paid to sales tactics and sales feedback for market sensing and customer contact.

Change-introduction. The Change-introduction organization includes the technical product-process definition groups who translate "what" into "how"; define the details of changes to be introduced into production; and provide the physical hardware, systems, information, training, and guidance to make the process work together in production. Suborganizations include specific project integration, project management, advanced technology development (on-going), production resources (tooling and facilities) and operational systems. This is the core of change-introduction—the technically creative concurrent product-process design effort.

Changes in the form of new product introductions, new processes, and new systems or facilities are implemented as projects utilizing planned, coordinated project teams. *Ideally a doable, relatively small number of discreet projects is in process at any one time with varying start/end dates for leveling workload and resource demands.* An overall project integration group coordinates scheduling and direct support. Because of the pivotal position of projects, this subject is discussed in more detail in a subsequent section.

A separate on-going advanced technology group develops and invents critical processes, materials, systems, and techniques, based on signals from tactical planning. The intention is to forecast technical need and derive solutions before committing to product design, so that risk is lowered and design cycles can be shortened. By making product design more predictable, new products can be introduced more at the tactically desirable timing, not at the "when we finish it" timing, which may be too late.

It is important that project responsibility continue until fully implemented results are in place, that is, until satisfied-customer reports are received by product service. The all too easy practice of considering that, if drawings are delivered to manufacturing, the product is "introduced," is not acceptable in this environment. Production's job is to produce, not to debug tools and equipment. One aid, to set the scene, is to design, procure, and put in place final test equipment very early in the cycle. Tight test parameters begin to establish and force the performance and quality levels for fabricated and purchased components as these are defined and tooled.

Production. Production is charged with converting sales orders into products in the least possible time cycle, given that product quality is fundamental, nonnegotiable, and must be continually improved. What is different here is the very close tie between selling (outside) and producing (inside). Past practices have permitted bureaucracy, delays, and the selling of products and services that could not reasonably be delivered with the production systems then in place. Short production cycles now required are forcing electronic links between customers and production scheduling decision points in supplying factories. For OEM deliveries, part-for-part synchronization of schedules in both supplier and user plants is setting the standard (see Chap. 6).

One firm having high-volume operations recently targeted reducing its basic product delivery cycle from 18 weeks to 2 weeks, and set about achieving this through stringent analysis and improvement of each step in the process. Time consumed in order processing, definition, scheduling, assembling, and delivering was examined, and teams were assigned with tough targets. Simultaneously, the product mix, tooling, modular product-line structure, and related physical factory facilities were modified to support shorter cycles. To date, progress has been very rewarding, illustrating again that concurrent product-process design is most effective when integrated with other business goals.

Selling and service. Selling and service is the customer contact arm. It is also responsible for maintaining close coordination with production. In practice, the interface between selling and service and production is sometimes turbulent and misunderstood. The payoff for improving this interface is great because it directly affects the way the business interfaces with the customer. One vision is that of a sales representative in a customer's office with, for example, his or her portable computer with all product and pricing data available on disk, so that a complete purchase order or release can be negotiated on the spot and then inserted into the plant's production schedule via modem and confirmed before the representative leaves the office. In principle, anything slower is not good enough.

Support activities. Support activities include all specialized functions that can be reasonably moved off the main line and done in parallel to the above activities. Support is not necessarily of lesser importance, but may be somewhat less time-critical. In general, support activities are intended to prepare, organize, backup, train, develop, and assure proper synchronized operation of the main line.

The most difficult support issue at all levels is training. The support functions must act as keepers of the knowledge, or at least brokers, to assure that any person in the main line is as fully prepared as the organization can afford. As technical complexity and the operating pace increases, the demand will justify a new order of competitive training of individuals: our engineer's ability versus their engineer, our manager's decision-making versus their manager. Continuity of operations preclude frequent turnover of talent. Training and retraining is therefore a necessity when turnover occurs.

Reducing New Product Introduction Time

New product introduction time may be defined as the time from when a conscious managerial decision is made to provide a new competitive product to the time when the new product is accepted by customers as being satisfactory in service (see Fig. 7.4). The purpose of stating new product introduction time this way is to mobilize and organize for action.

An obvious benefit of short development cycles is the business flexibility and ability to quickly respond to unexpected changes in the market and competition that it provides. We believe that there are many additional, but not so obvious, reasons for targeting short development cycles as an on-going strategy for the long haul. These reasons focus on the effective utilization of technology and continuous improvement of business practices and procedures.

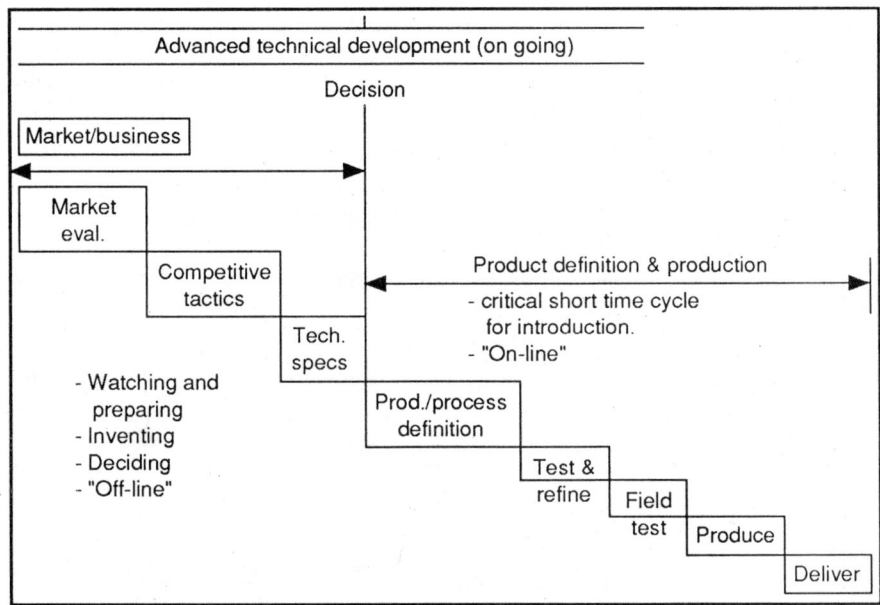

Figure 7.4 A generic new product introduction cycle combines these factors.

Up-to-date product technology

An immediate and direct benefit of urgency in product-process definitions is the earlier uncovering of problems and gaps in applicable technology. If this is coupled with an active experimental bias to breadboard,* mockup or simulate critical functions or processes as early as possible, then more time is available for problem solving.

A direct benefit of short times for product introduction is that customers' current needs can be satisfied sooner and probably more nearly matched. If delays persist, the product concept may be obsolete when finally offered or may suffer continual design changes and other expense in an effort to catch up. Also, the quicker the introduction, the more likely the product will satisfy customers' wants and, thus, command a premium price and healthy sales.

Use of the latest appropriate and reliable technology, by immediate application, is promoted by short development cycles. Good use is thus made of on-going, basic research and developments so that they can be ready when needed. Of course, good technical and managerial judgment is needed in selecting advanced versus reliable technology.

Deliberate shortening of available time promotes concentration on the business' core technology rather than dissipating resources on peripheral components (nuts and bolts) or on duplicating supplier's skills. By relying on a small number of proven suppliers or partners, their best talents can be incorporated early in the development. Then, internal proprietary development of the core technology can receive full resources. "Core" in this case is intended to include the focused technical capability at which the business is dedicated to be "the world's best" and to maintain superiority to all comers.

Indirect advantages of pressing for shorter introduction cycles include improved efficiency and elimination of marginal practices. This appears especially true for mature organizations that have built up many safeguards and internal procedures for transitory reasons but have not removed or even evaluated those practices for years.

For example, one firm had a variable speed control that used electromechanical components. Each different application rating required engineering to meet the customer's specification, resulting in long delivery cycles. Sales were declining and competition was severe. A new manager initiated a strong effort to redesign both the product and the process. Engineers were retrained and every person, including factory operators, received at least some training in basic computer programming. By innovative use of solid-state devices, variations in electrical

---

*A "breadboard" is an experimental electronic circuit constructed with separate components hand wired together on a specialized laboratory base (board) for evaluation and change of circuit values. It is useful in developing new prototypes that are then designed for production.

function were confined to software and firmware (programmable read-only memory chips). One immediate benefit was a drastic improvement in delivery cycle since an off-the-shelf standardized physical product could be customized very quickly and shipped "at the last minute" simply by installing appropriate memory values in the chips. Further results included major improvements in product cost, greatly reduced factory space, improved product quality and performance by more accurately controlling speeds, better reliability, and lower inventory. Market share improved significantly in spite of competition. Notably, the product, the process, the information systems, and the organization capability were *all* addressed at the same time—a winning combination, driven by a perceived need to introduce changes within short time cycles.

Contributed value

Internal cycle-time evaluation can be quite revealing by plotting where the time goes in detail and examining each action for the value it contributes to the final product. Even asking the question, "what do we consider to be contributed value?" is very educational and usually leads to more pragmatism and changes in operations. "Value" here refers to operational activities that result in establishing and/or enhancing the performance, quality, or delivery of the physical product. For example, "expediting" does not contribute value.

Problems that remain unsolved and "worked around" for long periods may be addressed bluntly and improved by the criterion of drastic shortening of development cycles. One example involved shortages of development parts for a high-technology product line. Since needed parts were either individually unique, or modifications of scarce parts, the shortages for test and evaluation of new or changed designs significantly extended development times. By examining existing procedures under the harsh light of meeting competition on a very short development schedule, barriers were identified and opportunities pursued. In addition to identifying long-lead items for castings, forgings, and so forth, a calculated program of variations in material dimensions for early development parts was defined to cover a range of possible needs. The increased cost of certain forgings was small compared to the program benefits of earlier availability of development prototypes and components for test runs.

For development testing, in this example, extensive instrumentation was attached to component parts within the product. Previously, installation of the instrumentation and associated modification of newly designed component parts was performed in the laboratory after receipt of the development parts. This was a very time-consuming process. By planning specific parts for needed instrumentation during early design phases, the modification and test time was significantly shortened and productivity was improved. However, these tactics probably would not have taken place without intense effort to get the job done quicker. *It is notable that many of the best time-saving ideas, observations, and solutions come from hands-on operators and others immersed in the work, not from remote technical or management people.*

Another indirect benefit of shorter cycles is simply that less time and fewer dollars are consumed on a given project by eliminating the marginally useful practices, that is, by doing only what is really necessary as opposed to what is comfortable or traditional. For example, it is traditional to partially machine in large batches the common dimensions of certain similar parts (for example, shafts), store them, then take parts one-by-one from stores and machine special dimensions. By coordinating the design of the parts with the tooling design, the parts can be machined completely by selection of modular quick setup tooling. This eliminates the entire storage and retrieval operation providing both cycle and cost reductions.

Although difficult to define or measure, an increased sense of urgency can also lead to more innovative technical solutions since small ideas don't have enough impact to achieve targeted results. Computer industry developers appear to have routinely incorporated this urgency into their product launch practices where competitive survival demands quick introduction of the latest available technology. A typical first question is "How can I eliminate at least 50 percent of this activity while providing the same or better result?" This is proving quite useful in office procedures and practices including CAD/CAM, configuration control, and other information handling.

The Project Approach

To effectively support the introduction of concurrent product-process changes, a project approach is a virtual necessity. A project in this sense is defined as a specific task-oriented internal group of people organized and dedicated to achieving a single (usually major) objective then disbanding. Typical tasks might be "design and produce a new model X product to be introduced at IMTS trade show in September '91," or "develop and install a wire-EDM process in production that engineering can use for individual custom-contoured cams, thus eliminating entire G–96 control box," or "develop an embedded computer system architecture and protocol as common interface for peripherals A, B, and C."

Currently, most significant changes are implemented through such projects, yet we do not always organize them to the fullest advantage. One might ask, why not recognize the effectiveness of focused project operations, design the project organization to be maximally effective, and use it as the building block for operations. This is normal for certain high-tech systems businesses where it is necessary to be responsive in highly competitive situations.

Packaging change-introduction into discrete projects helps center the organization's alignment on change-introduction projects and the exploitation of these changes in sales, production, and delivery. The more rapid the changes in the business environment become, the more necessary this focus on change-introduction becomes and the more critical it is to the business to have effective project management of change.

Evaluation of varied project experience at GE and other firms provides many insights into the factors that contribute to a successful change-introduction project. The overwhelming lesson learned is that *how* projects are conducted is a vital issue. Although much has been published about the internal administration and operation of projects, in practice, certain key considerations are frequently overlooked or forgotten. In the following sections, we review critical success factors that have been distilled from GE project experience.

Maintain a specific product and market segment focus

Focus on a specific product, intended to serve a specific market segment for vital strategic and tactical reasons. It is critical to have well-defined specifications of what the final product must perform for the customer to provide the competitive advantage being sought.

Many product design programs are initiated for internal reasons (margin, cost reductions, and so forth) and result in either the wrong product, an uncompetitive one, or one which is too late. The transition from the market's judged desires to product-process definition and detailed engineering is key and must be very carefully managed. How can this critical transition be improved? *Successful projects appear to have a strong customer thread.* That is, the decision to begin a project is *preceded* by a very careful evaluation from a business and market viewpoint of *what* to design (see Fig. 7.4). The evaluation encompasses an analysis of *what* the envelope of allowable cost, timing, appearance, and promotion includes. It does not necessarily address at this point the issues of *how* to design the product. A key feature of the evaluation is that it includes the very best thinking from all aspects of the business including technical, financial, talent resources, capacity, and others.

Project leader

In post-mortem examinations of projects, *the single most essential factor is often found to be the top level leader.* Characteristics emerging from successful project leaders include those below. Where applicable, illustrations are drawn from GE projects to emphasize that these are *people,* not just a set of capabilities.

1. *Respected both for business judgment and for technical expertise:* In the earlier example of upgrading the variable speed control, the newly appointed general manager had both broad design experience in the more advanced technology and also enough business insight to envision how the competitive position could be greatly improved. His observation was that the business had been "knocking on the same doors" for too long. His views were not initially accepted within the organization, but as he built up the new application technology and brought in new education and training, respect for the new approach and for him personally grew rapidly. When sales and income began to improve, the organization's motivation and support became a strong and continuing asset.

2. *Total dedication, to the point of obsession, with the particular project:* In 1982, at one firm's consumer goods divisions, it was apparent that an improved production facility was needed. An appointed manager was charged with the task and set about designing the product and the processes simultaneously as an integrated and coordinated system. For example, instead of a large and complex fabricated metal component, this manager conceived of a one-piece molded plastic component using new high-strength material that could be automated. This and similar "radical" ideas eventually led to a product that achieved excellent reliability and customer acceptance. Project results have included improved market share and a production facility that has become a national show place for what clever automation can accomplish. Accomplishing this, however, took nearly five years and involved many "agonizing reappraisals" by the project team. Without the intense personal dedication and perseverance of the project manager, the project would not have succeeded.

3. *Charisma:* Hard to define but conveying a strong impression to others of knowing the direction to go, welcoming other travelers, and convincing them to follow. In large organizations, charismatic people have a difficult path. An outstanding example was the individual who

managed a major high-tech business for many years of growth. This person's unusual energy, technical insights, interest in people and products—and always better products—inspired a following that bordered on fanatical. This inspiration built the first innovative product, developed its derivatives, and carried it from a project into a base for world leadership. This was accomplished in spite of the demanding nature of the product and the most aggressive competition imaginable.

4. *Stamina to continue with the project until all goals are met:* Leading change is seldom an agreeable and accepted position. For these managers to continue on the path of their projects and bring together the vision, the technology, the investment, the union, and the customers' desires until all objectives were satisfied illustrates the level of stamina required.

5. *Positioned to have authority to cause instant action without having to negotiate, arbitrate, and plead for support:* In the days of pooled resources, of matrix management, of straight line and dotted line reporting relationships, and so forth, it is a challenge anywhere in the organization to acquire authority and then to use it when necessary. When in trouble, or when a clear direction and stability are called for, a recognized and decisive authority is required.

Project definition, scope, and timing

Many other factors can be cited as contributing to a project's success. Based on experience at GE and other firms, the following factors appear to be particularly relevant to success in project definition, scope, and timing:

1. Build on existing core knowledge, that is, what the organization does best, wants to continuously improve, and urgently intends to maintain as superiority over competitors. Develop a few superior partners and suppliers to provide their own core expertise and components and rely on them for noncore technology. Don't waste talent and time designing components that can be bought with good quality on the open market.

2. Define the business and market targets first and in such a way that, if these are met, the project will be successful. These goals will normally involve considerable stretch. A five-year operating plan extending back five years and forward five years will enhance perspective and show up discontinuities in plans. Sudden major improvements in forecasts are unlikely to actually happen.

3. Use front-edge-of-the-art technology in scoping and targeting product performance and the enabling process capabilities. Of these,

the process technologies are frequently *more* significant than the product configuration and deserve more attention than they usually get. Further, the processes are more easily guarded internally, are more difficult for competitors to "reverse-engineer," and have profound impact on product cost and quality.

4. Progressively define the product specifications within the business and market constraints in terms of customer values. The concept is inherent, for example, in "QFD" (Quality Function Development). Perform detailed analysis of what constitutes acceptability to the customer and of how much it will cost to provide those characteristics in the product-processes. Since this analysis takes time and effort, the more aggressive efforts are usually made where (1) customer usage of the product is frequent and intense, (2) product quality differentiation from competitors is vital, and (3) a clearer criterion for process and tooling perfection is needed. This has in the past meant consumer products but should also include industrial products. Integrated analysis of product specifications is used within GE and other firms for products as diverse as major appliances, industrial equipment, and high-technology, quick-development products.

5. Limit the number of product projects to the very few, most critical items that provide the greatest expected business return, and exert all efforts to avoid scheduling critical peak efforts that overlap in a limited resource such as design analysis or tooling.

6. Provide clearest possible design targets of product performance specifications that are critical and less definition of the materials, processes, and configuration, thus, *allowing more creativity* in the "how-to" of the product-process definition.

7. Include full product life-cycle planning to relate expected development costs, sales volume, financial plans and tracking, anticipated competition moves, anticipated changes in technology and material costs, and so forth. Certainly these will change in due course, but a viable base plan is needed for guidance and for updating.

8. Provide clear and complete *business* plans (see Chap. 6) and scope to the teams charged with introducing change, that is, product-process design, facilitation, and production. Involve cross-functional team members from the outset.

The project team

A serious project deserves that serious attention be paid to the balance of knowledge of team members right from the start. Any gaps will have to be filled later, which is slower and more expensive than filling them at

TABLE 7.1 Integrated Team Activity During Phases of a New-Product Project

	CONCEPT	SPEC
	Phases	
Team knowledge	Tactics & viability	Focus & impact
	Team Composition	
Business and finance management	• Bus. plan & tactics • Competitive eval. • Niche ident. • Fin. viability • Invest/divest	• Vol./margin est. • ROI, ROS, ROA • Cash flow plans • Risk analysis • Resource allocation • Concept judged viab.
Marketing	• Product opportunity • Market research • Sales feedback • Forecast • Concepts, prices • Trade-offs	• Concept/niche vol./demand • Features/quality • Product range • Perform. range • Market competit.
Product-Process Eng'g.	• Tech. trends • Enabling processes • Prod.-process struct. • Cost reduction • Perform. upgrade	• Concept definit. • Alternates • Cost est. progr. • Prod. cost est.
Prod'n Systems and Facilities	• Logistics, OSB • Supplier relat. • Systems/ed. • Facil. capacity • Costs. M/B	• Cost of changes • Facil. impact • Other resources impact
Production	• Capacity/loading • Prod. mix impact • Resources impact • Training req'd. • Operations • Investment	• Program plans • Planning est's.
Sales and Customer Service	• Sales/impact • Order serv. impact • Distribution need • Cust. serv. impact • Surveys	• Field data gather • Cost/price eval.

the outset. Even in early stages of planning competitive tactics, there are questions relating to each operation in the business (see Table 7.1). For example, manufacturing questions, which tend to be postponed to later, offer both opportunities (process and cost leadership) and constraints (over- or undercapacity, supplier issues), which should be recognized. The core team should include expertise from most areas of the business.

TABLE 7.1 (*Continued*)

DESIGN	CERTIFIED PRODUCT	PROD'N RELEASE
Design	Verify	Facilitize
	Team Composition	
• Bus. decision: accept program • Monitor progress • Dev. auth. budget • Mid-course correct • Vertical integration	• Assure continued business viability • Monitor adherence • Update planning • Verify measurements	• Provide investment • Monitor external factors • Assure internal integration
• Product application structure • Trade-offs • Alternatives • Product planning integration	• Prepare field trials • Monitor applications • Define mod's. • Assure accept. to customer needs	• Plan intro. • Assure readiness • Guard key info.
• Product structure • Process interact. • Product design • Dwgs./specs. • Config. control • Req's./sample pln.	• Prototypes/models • Field evals. • ECN's, PCN's • Process perfection • Tooling/methods • Cost	• Assure qual. defin. • Assure qual. perf. • Prod.-process debug
• System plans • Facil. plans • Long lead changes • Negot. suppliers • Mat'l. avail/cost • Design reviews	• Supplier qual. • Facil. details • Long lead facil. • Training • Investmt. monitor • Quality syst.	• Install syst/data • Comm. syst. on line • Debug
• Samples • Process verify • Resource impact • Long lead items • Intro. schedule • Design reviews	• Rearrangements • Invest. monitor • Start-up plans • Quality plans	• Install/debug • Train • Pilot • Rate ramp-up plan in place
• Design reviews • Eval. customer react.	• Prototype eval. • Customer react. • Customer preparation • Plan customer serv.	• Adv. & sales prom. • Customer contact • Distrib. prep. • Pre-sale act. • Introduction • Customer service

Team expertise may be supplemented in phases where workload is heavy, but note that at all time periods, maintaining perspective requires at least a reference base of each knowledge area.

The project manager and other key team people must have full-time assignment to the project, direct-line reporting, and no part-time or diversionary duties. In matrix organizations where project staff are full-time reports to their functions and only dotted-line to the project, the project frequently suffers unclear direction, conflicting orders, and missed schedules, as well as less-than-effective use of expensive talent.

Communication

Given the complicated issues, effective communication both within the project team and to supporting operations is critical and yet, communication never seems to be really adequate. GE experience has shown the following aids to be useful in helping to improve effective communication:

1. Core talent people stay with the project from its inception and know each other's views, contributions, and work status. Few people are added later so that less catching up is needed.
2. Team members sit together, work together, attack problems together, and devise solutions together. Geographical distance, whether within plants or between cities, is a major barrier to effectiveness.
3. Make joint plans, hold joint planning and status meetings, use common schedule formats.
4. Drive toward maximum use of electronics for all interaction, including computer E-Mail, messages, networks, multi-access databases, progressive updating of data, and so forth. The present rapid advances in systems technology as applied to internal organizational communication are providing continual improvements.
5. Push for physical demonstration of progress, not progress reports. Simplify communication: either it's there or it's not; either it works or it doesn't.
6. Package work so that most of the communication is within the work group. Then the useful communication to others consists largely of "the task is, or is not, complete for input to the next work group."

Advisory council

When totally immersed in the work of development, it is easy to lose perspective. One stabilizing method is to appoint an Advisory Council or Steering Committee, usually made up of senior staff, to periodically review project progress and evaluate status (see Fig. 7.5). To be effective, it is important that the Council provides a detached viewpoint and that it has enough influence to reallocate resources if a given task falls behind or goes too far ahead.

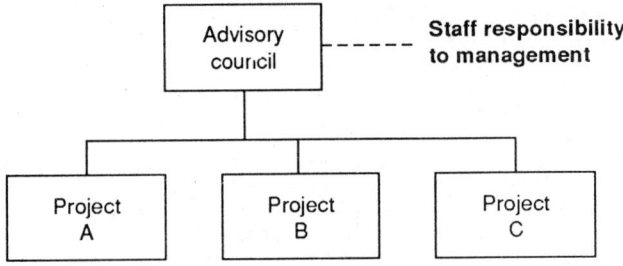

Figure 7.5 Advisory council structure.

During recent vigorous upgrading efforts of one firm's high-volume business, a total of 11 major projects and many more sub-projects addressed nearly every aspect of the business. This included 17 different plant locations producing many variations of the basic products. The advisory council reviewed each project's status frequently, every one to four weeks depending on urgency, and the council's judgment and actions, defacto, assumed the major role in operating the business. This experience has been highly successful and contributes both insights and further questions. For instance, if this approach works so well for introducing change, might it thus be a model for on-going operations during a period when continual change is needed for survival?

The Salisbury Project

Do the organizational and project concepts outlined in this chapter work? The General Electric operation that produces panel board components for electrical distribution equipment undertook the redesign of their products and production processes with the objective of improving service to their sales and distribution network by simplifying internal operations. This project was carried out using an integrated team, including both technical and marketing expertise, from the beginning, in the manner we have described in this chapter.

The Salisbury results

Major results to date are summarized in Table 7.2 and include the following:

1. The number of parts required was reduced from 28,000 to 1,275 by developing modular component designs adaptable to multiple uses, by product structuring using the team approach, and by using other Design for Assembly concepts.

TABLE 7.2 GE Salisbury, N.C., Lighting Panel Plant Design for Assembly and Flexible Automation Team—Work Results

Original	GE Salisbury
6 plants	1 plant
28,000 parts	1275 parts (maintaining 40,000 customer options)
Backlog of 2 months	backlog of 2 days
	product models change 12 times per day

2. "Requisition" engineering (hands-on engineering of each customer order) was eliminated while more features and versatility were designed into the product itself using a modular building block approach. With the new design, 40,000 customer options are possible through various combinations of the building blocks.

3. Overall cost was reduced by a double digit percentage while achieving significantly higher quality. Direct labor in the product was significantly reduced accordingly, as was routine salary effort. Computer-based simulations of processes and systems enabled major operating insights into process costs. The ROI on investment in equipment and tooling met stringent criteria.

4. The cycle time for shipping a customer order was reduced from two weeks to three days, two of which are spent getting the order from the customer to the appropriate starting point in production. Product models are reportedly changed a dozen times or more a day and the plant has improved productivity by 250 percent (*Business Week,* 1989). Backlogs in the plant have gone from two months to two days (*Fortune,* 1989).

5. Market share was increased by 3 percent while the project was being installed with the expectation of further improvement.

6. Production scattered in six locations was consolidated for efficiency.

7. Personnel turnover in the plant has gone from 15 to 6 percent after the first year of using the flexible automation system. The number of worker hours per distribution board is reported to have been reduced by two-thirds (*New York Times,* 1988).

The Salisbury solution: The story behind the story*

Objective: In the mid-1980s, GE Electrical Distribution Operations wanted to be the market leader.

---

*Portions of the following section are reprinted with permission from Stephen L. Harris, "Salisbury Solution," *GE Monogram*, vol. 66, Winter 1988, pp 26–29.

Problem: To reach this goal it knew it had to clear several obstacles, the toughest of which was to move the manufacture of its product to a new location (Salisbury, N.C.) and then retrain the already established work force there.

Solution: The panelboard was redesigned to allow for automated manufacturing. The traditional way of running a plant was scrapped in favor of an organizational redesign in which the "self-directed work force" is a part.

Is the solution working? GE's *Monogram,* a company publication, visited production employees at the Salisbury Plant to find out. The employees' comments are shown below. The degree of acceptance shown at the plant operating level highlights the potential for improving basic work flow and unleashing people's involvement through carefully designed work teams. This same enthusiasm can, and should, extend throughout the business from product planning through product-process technology and into production operations.

Plant Manager: "The reality was that the product was in serious trouble. Significant changes had to be made quickly in order to effectively compete in the marketplace. As a result of the new product, flexible automation, and streamlined systems, we knew that we could significantly reduce cost and improve service. However, we needed a human resource plan that would complement these approaches so we could develop a work force with the ability and confidence to make the day-to-day, hour-to-hour decisions necessary. The first step was to remove first-line supervision and reorganize into self-directed work teams. This resulted in a reduction from four to two layers of management and enhanced communication. These work teams assumed responsibility for such things as production, quality, resource allocation, and plant policies and practices. To help cultivate the technical and social skills necessary to function effectively in this new environment, a pay-for-knowledge system was implemented. Simply stated, this approach rewards individuals for the number, kind, and depth of skills they develop rather than the particular tasks they perform. The difference in what this approach has meant to our business was best stated by one of our people, Opal Parnell, when she said, 'Today, we *are* the business.' In Salisbury, self-directed work force is not a program, it is a philosophy."

Tim Deal, Production Control: "I like the fact that it's self-directed here. I think people like to be in control of their own destiny. It gives us a chance to get involved. Of course, not everyone has accepted that responsibility. But it's coming along. It's frustrating at times. We're trying to solve problems and it may not seem like we're making progress. It's more like taking two or three steps forward and one step back. But that gives us an appreciation of the problems and responsibilities our supervisors faced under the old system. Now we don't have the benefit of the go-between. We've got to solve our own problems."

Jim Lisk, Materials Manager: "I have been here since 1975, and I was apprehensive about all the changes. I foresaw problems because the magnitude of the change was so tremendous. We changed everything but the four walls of the plant, including the way people were supervised. We jumped in feet first with no specific implementation schedule for the self-directed work-force concept. We didn't have a timetable because we were dealing with a wide variety of traditional attitudes. Some grasped the concept faster than others. But is was exciting, it was new, and it was challenging. At first I was a skeptic, but I became more comfortable after experiencing the participative environment. Communicating the vision was one of the most frustrating and difficult challenges. One of the things I learned is that you have to have patience. We struggled with the decisions about when to get involved and when to back off. People had to learn to solve problems for themselves. So the dilemma was always: How much leadership and control do we provide and when? I have found that this self-directed work-force concept brings out the best in people. It not only provides growth opportunities in our business lives, but growth in our personal lives as well. More than anything, I have found that if you display confidence and trust, people will surprise you."

Summary

Management's success inevitably depends on making the most effective use of expensive resources such as people and facilities to increase the value of material passing through the enterprise. Ideally, low-cost material would be designed and fabricated into high-sales value products, using minimal cost processes and facilities, both during development and production. Under intense competition, business margins become small and all resources must contribute to their maximum to provide product value. The system must work at maximum effectiveness and efficiency.

Maximum effectiveness and efficiency can be achieved by focusing the organization's internal operations on change as the dominant driver with design cycle reduction the chief objective. Providing the tactically competitive product at the right time requires that the entire organization work together, with each element contributing what is needed when it is needed. Ultimately, all activities within a business have but one purpose and that is to serve a customer so effectively that net profit is made.

The themes developed throughout this chapter can be summarized as the following recommendations:

1. Emphasize the initial marketing and business planning of the product to be introduced. Apply the best tactical minds to defining exactly what product will satisfy the selected market segment and preempt competition. Specify the product in terms meaningful to project management for introduction. Plan ahead, do inventing and development off-line so that required technology is ready and available when needed.
2. Organize the new product introduction process, from start to finish, along work flow lines, not along functional hierarchy lines. Condense the organization into groups arranged in the sequence of work flow through the business. Perfect the output of each group as input to the next group. Reduce low-value work, using short product introduction cycle as the criterion.
3. Organize new product projects to have full autonomy and full support and also provide careful progress monitoring by an advisory council.
4. Focus product planning, product design, and process design into one cohesive working group with minimal distractions. Respect the intensity needed to blend together the factors leading to a product which satisfies the customer, is within investment constraints, and is practical and deliverable on time.
5. By continuing detailed analysis and action, reduce the new product time cycle to less than the competitor's time cycle. Management flexibility is enhanced and the elimination of low value work improves effectiveness.

References

Business Week, "The Best-Engineered Part is No Part at All," May 8, 1989.
Fortune, February 13, 1989.
New York Times, October 17, 1988.

Chapter

8

IBM Corporation: Early Manufacturing Involvement (EMI)

William H. Monsen

Introduction

The keys to competitiveness in the computer industry are product innovations and product cycle reductions. It has also become very clear that these items are not mutually exclusive. Product innovations, if not brought to fruition in timely product announcements, can render a company "noncompetitive" in a matter of a few years. The total product cycle—design to manufacture—must be improved from the traditional serial process to a total interactive process.

In IBM, we have recognized that early manufacturing involvement (EMI) in the product design is key to reducing the total product cycle. We can no longer tolerate three design iterations—one for function, the second for manufacturability, and the third for cost reduction. Designs of the future must include function, manufacturability, cost, and quality in one pass. Technology now changes so frequently that a series of design iterations at a given technology level cannot be completed prior to the next technology innovation.

IBM Rochester, Minnesota

IBM in Rochester has been using the concept of Early Manufacturing Involvement to integrate design and manufacturing in the product cycle of midrange computer systems. Examples of product cycle improvements using EMI will be reviewed in this chapter.

The IBM facility at Rochester, Minnesota, was established in 1956, with 175 people working in a leased 50,000 ft² facility. A new 518,000 ft² manufacturing plant was dedicated in 1958. In 1961, a development laboratory was added; and in 1985, IBM Rochester expanded by an additional ½-million ft² into two new buildings.

Currently the Rochester site consists of 586 acres. The IBM Rochester manufacturing plant and development laboratory occupy more than 3.6 million ft² of owned and leased space. IBM Rochester is the largest IBM facility in the world under one contiguous roof. Current employment is about 7,000 full-time, regular employees.

The IBM facility in Rochester is the largest manufacturer in Minnesota outside the metropolitan Twin Cities area. The Rochester site is also one of the largest intermediate data processing systems development and manufacturing facilities in the world.

Rochester is the principal site of the Application Business Systems line of business, one of seven units that compose IBM United States. Rochester is responsible for worldwide hardware and software development and U.S. manufacturing of midrange computers including the Application System/400, (AS/400), the System/36, (S/36), and the System/38, (S/38). The site also has development and manufacturing responsibility for low-end direct access storage devices (DASD).

Midrange computers, costing between $10,000 and $1 million, make up a $26 billion market worldwide, encompassing an estimated 13 million potential customers. Midrange computer technologies involve electronic circuitry and software, including both operating systems that control the internal workings of the computer and application programs that help customers run their businesses. The midrange market, in which the Rochester site competes, is extremely competitive; continual improvements in all business processes are a necessity.

Early Manufacturing Involvement (EMI)

Let's get back to EMI in the product cycle of computer systems and define the term. An EMI process involves product design personnel and manufacturing personnel working in close partnership throughout the design process to optimize function, manu-facturability, cost, and quality in a one-pass design.

The word "manufacturing" in EMI includes all typical manufacturing plant functions—purchasing, production control, industrial engineering, quality engineering, manufacturing engineering, and manufacturing line personnel. The EMI concept is not at all complex. Cooperation and timeliness are key elements.

Throughout the product development cycle of a new computer system, there are key reviews and checkpoints that become part of a comprehensive product plan. This product plan also includes tests and

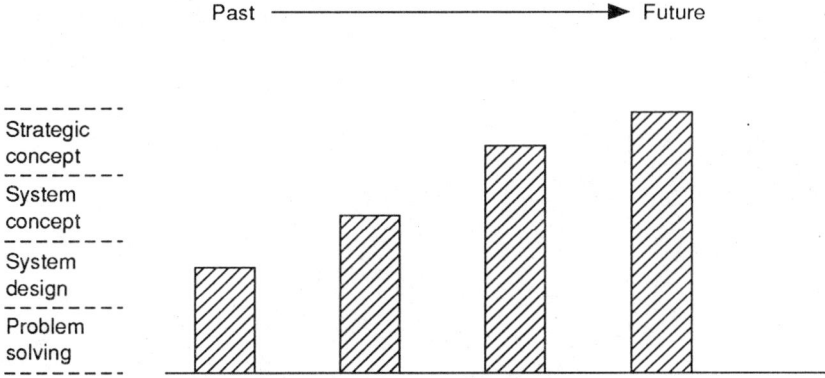

Figure 8.1 Level of manufacturing involvement in design.

evaluations with specific criteria to be met. Management and technical reviews assure the requirements of design, manufacturing, quality, and marketing/service have been met. Representatives of these functions meet regularly to identify concerns and resolve issues. The EMI concept is a discipline within manufacturing that emphasizes continuous interaction in the design process during this overall product cycle. This level of manufacturing involvement in the design process (see Fig. 8.1) has steadily increased over time.

Prior to EMI, completed designs were sent to manufacturing for review before being released as official documents. At this late stage in the design, minor modifications could be made, but changes requiring redesign were avoided because of time constraints. Design changes suggested by manufacturing usually focused on ease of manufacture, reducing part cost, and reducing tooling cost. With emphasis on early involvement (see Fig. 8.2), manufacturing can now focus on hardware concepts, logistics, commonality, software, affordability, and quality.

Technology improvements in design and manufacturing hardware and software have also assured product cycle reductions by eliminating paperwork along with its associated time delays and occasional errors. Computer Aided Design (CAD), Electronic Data Interchange (EDI), and Computer Integrated Manufacturing (CIM) have had a profound effect on design-manufacturing integration. (These functions have been the subject of considerable technical documentation and are not dealt with here.)

One-Pass Design

What occurs with EMI and why is it effective? At the concept level in a product or part design, many different considerations can be incorporated without major impact to total design time. A cooperative ef-

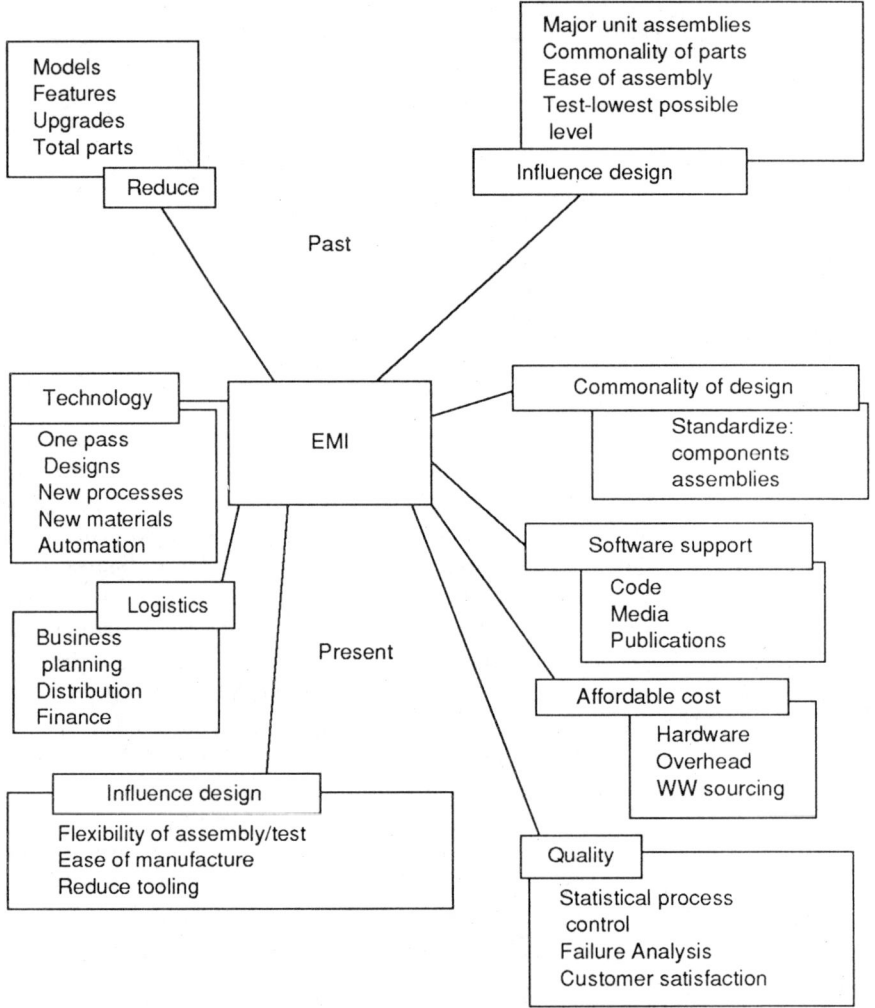

Figure 8.2 EMI design influence.

fort is needed and timeliness is critical. *The goal is a one-pass design.* After the initial design concept, changes are usually difficult to incorporate and are often ineffective compromises. They typically lengthen the product cycle. Early partnership between design and manufacturing has proven very effective in providing product and process design cycle improvements. Reductions in the overall product cycle of up to 50 percent have been achieved. Within IBM, the manufacturing engineer has typically been the EMI team leader. Other team members include representatives from purchasing, production control, industrial engineering, quality, and the manufacturing line. The key EMI goals are to (1) reduce the development/release cycle time, (2) reduce costs and assure quality, and (3) design for manufacturability.

Early Manufacturing Involvement teams are organized while the product's design is being formulated. These teams interact with the product designers to identify each organization's needs relative to the design. The manufacturing engineer is the focal point for the design groups and coordinates this interaction. Weekly meetings review progress and identify both product and process focus areas.

EMI Product Focus

The Application System/400, (AS/400), computer system is a good example of many excellent EMI successes. One of these is assembly of cables into this computer system. Early cabling concepts on the Model 9404 used multiple cables, many fasteners, various terminal blocks, and assorted electrical connections. Designing a plastic carrier to replace many of these parts (see Fig. 8.3) significantly improved manufacturability. Multiple cables were designed for assembly onto this carrier, which could be built and tested by a supplier. This allowed the unit to be delivered and stocked as a single component on the assembly line and, thus, reflected improvements in inventory, assembly time, and just-in-time delivery. In this example, manufacturing engi-

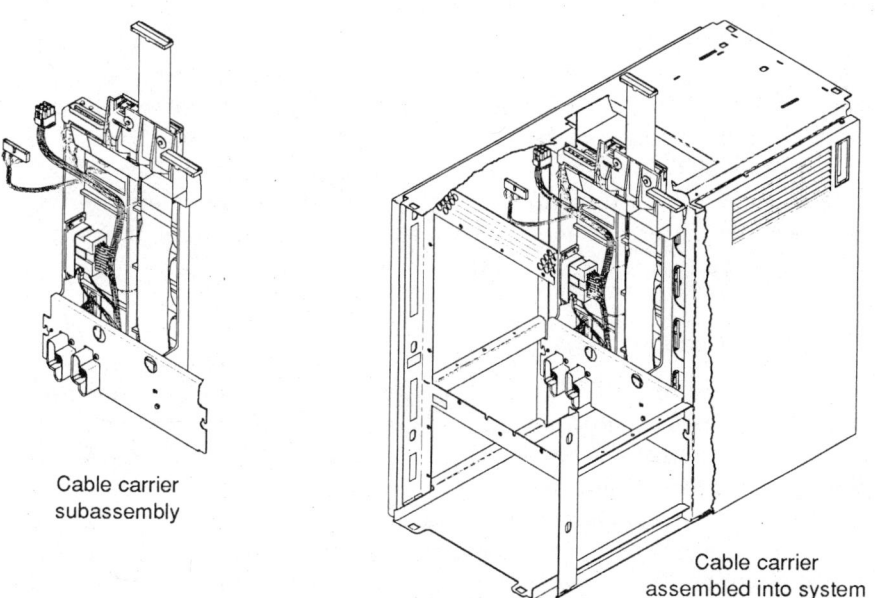

Cable carrier subassembly

Cable carrier assembled into system

Figure 8.3 Modular cable carrier subassembly as a unit and installed into a computer system.

neering, quality engineering, and manufacturing line personnel worked together to define the cable carrier. They then presented their proposal to the design group. The designer reviewed the concept for functional needs and incorporated this unit into the overall design. Manufacturing input during the design stage clearly enhanced the assembly flexibility and reduced the capital investment in the manufacturing process. Figure 8.3 depicts a modular cable carrier subassembly as a unit and installed into a computer system.

Another example of EMI product focus is system testing. To assure an extremely high-quality level of product performance on a short product cycle, card and system testing was done at the prototype design level on the AS/400. Manufacturing engineering designed simplified early test equipment. This equipment, set up in the design laboratory, simulated the manufacturing line test equipment. Early card prototypes were tested on this equipment to provide performance information for the card designers. A margin of safety, or guardband, was established in this testing to ensure that follow-on production orders of parts, obtained from multiple sources and tested on manufacturing test equipment, would operate properly through the entire system specification range. Early detection of problems allowed time to modify designs and improve manufacturing quality before volume production began. The simplified testing in card and system manufacturing improved the quality of the assembled product at a significant cost savings.

A final example of product focus is design for modular assembly. Designing plugable subassemblies increases the capability to customize a product, with little or no increase in assembly complexity. Uniform design of subassemblies allows use of standardized parts storage. If customer model demands shift over time, the assembly floor layout can remain relatively unchanged. Setup time between models is eliminated because different models use the same basic equipment and assembly operations.

EMI Process Focus

An EMI process focus includes emphasis on all manufacturing processes. As product design information becomes available, various specialty manufacturing engineers review the designs, evaluate process impacts, and recommend design improvements. Suppliers are given early views of part designs; their recommendations are fed back to product designers. Each part is reviewed for manufacturability, delivery lead time, tooling requirements, packaging/shipping expense, and commonality. The results and recommendations of these reviews are continually rolled up and fed back to product design groups for final design consideration. In this manner, manufacturing process evolution coincides with product design evolution.

Early Manufacturing Involvement, with product and process focus, targets the following results:

1. Product development cycle reduction (see Fig. 8.4):
   - *Lower (faster) manufacturing learning curves because of early design involvement.*
   - *Minimal redesign/retest for manufacturability.*
   - *Early tooling/processing capability established with manufacturing making both the prototype parts and the production parts.*
   - *Cross-training between design engineers and manufacturing engineers. Manufacturing engineers with better visibility to design processes and design engineers with insights into manufacturing processes.*

Past

Development cycle	
• Limited early mfg. involvement	• New ASM line/equip. for each prod.
• No cross training	• Cumbersome test equipment
• Redundancy dev. ◄►mfg.	• Assemblies were labor intensive

Present

Development cycle	
• Manufacturing engineers in development writing code	• Statistical process control
	• Common testers (DEV. and mfg.)
• Manufacturing design	• Prototype parts from mfg.
• Improved quality levels	
• Multiproduct/flexible assembly lines	

Future

Dev.cycle

Figure 8.4 Product development cycle reductions.

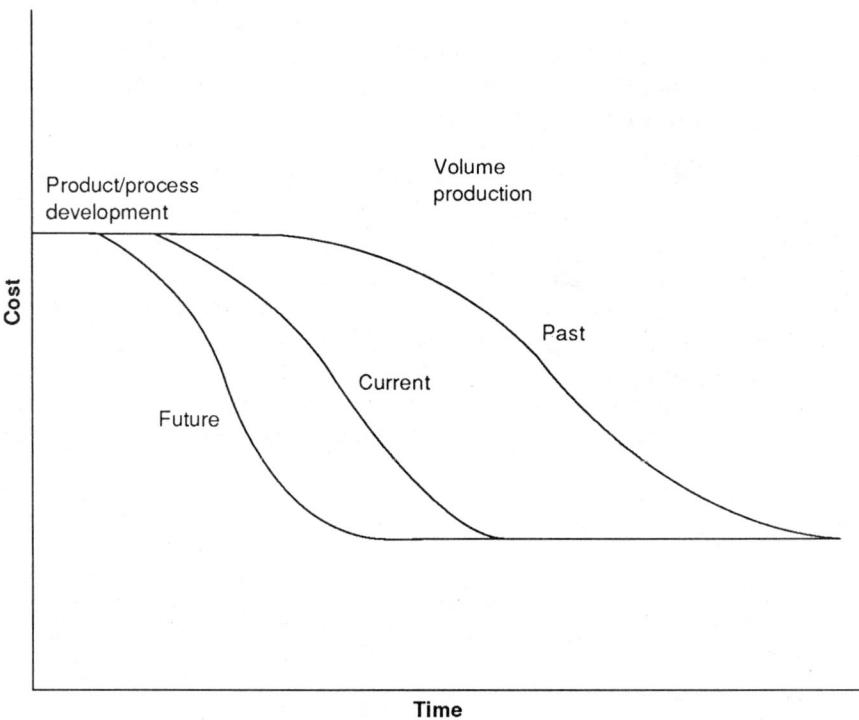

Figure 8.5 Use of EMI to reach ultimate manufacturing cost as early as possible.

2. Reduced costs (see Fig. 8.5) and optimum quality (see Fig. 8.6):
   - Early emphasis on all cost issues. Hardware costs, overhead costs, sourcing, and continuous flow manufacturing become balanced against projected revenue before the design is "frozen."
   - Latest processing techniques are incorporated considering function, cost, and quality.
   - A manufacturing experience base with functional understanding established earlier for continuing product support.
   - Quality goals established early to drive improvements over previous products.

3. Manufacturability:
   - Design flexibility for assembly and test processes incorporated at design concept stage.
   - Design influenced for commonality to allow reuse of existing equipment and processes.
   - Early manufacture or purchase decisions allow designing the part for a known manufacturer or technology.

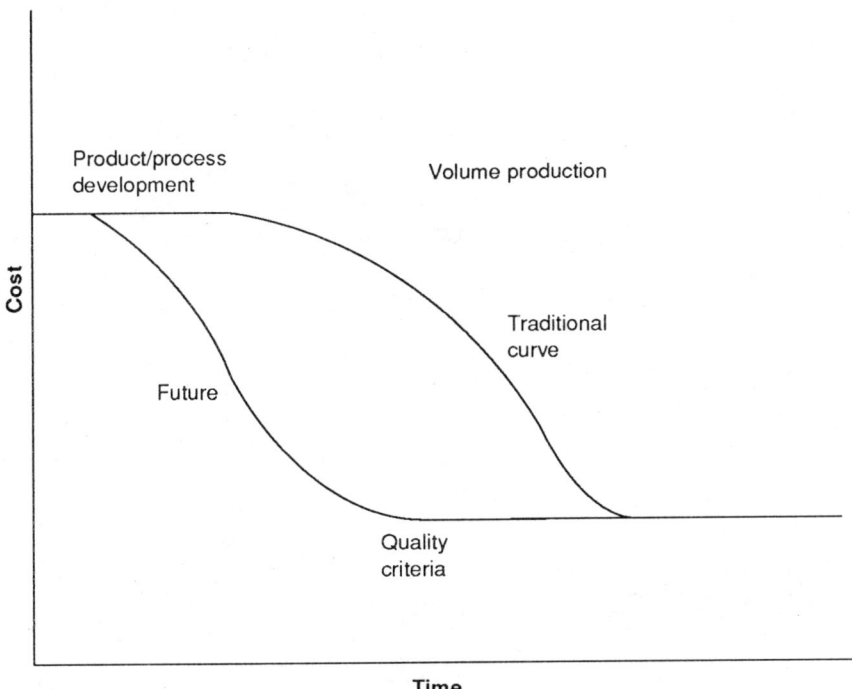

Figure 8.6 Use of EMI to reach ultimate quality criteria with early failure analysis.

Manufacturing Designers

The development and manufacturing processes have been merged to become one common, interactive process. This merger has advanced to the point where manufacturing engineering is doing the mechanical design/development work on low-end computer systems.

The IBM System/36, Model 5363, announced in October 1987 accomplished several "firsts" for IBM. *This low-cost system was brought from concept to delivery in only 11 months. The development cycle was reduced more than 50 percent using manufacturing engineering as the mechanical system design team.* This was design-manufacturing integration in its purest form!

This mechanical design team, composed of manufacturing engineers and procurement engineers, had *CADAM* graphics system equipment experience, but had not done product design work prior to this project. Their backgrounds were a variety of mechanical manufacturing skills from tool and model maker training to mechanical engineering. Their familiarity with *CADAM* was from using this system to review designs of product design groups. Figure 8.7 depicts the mechanical hardware designed for this system.

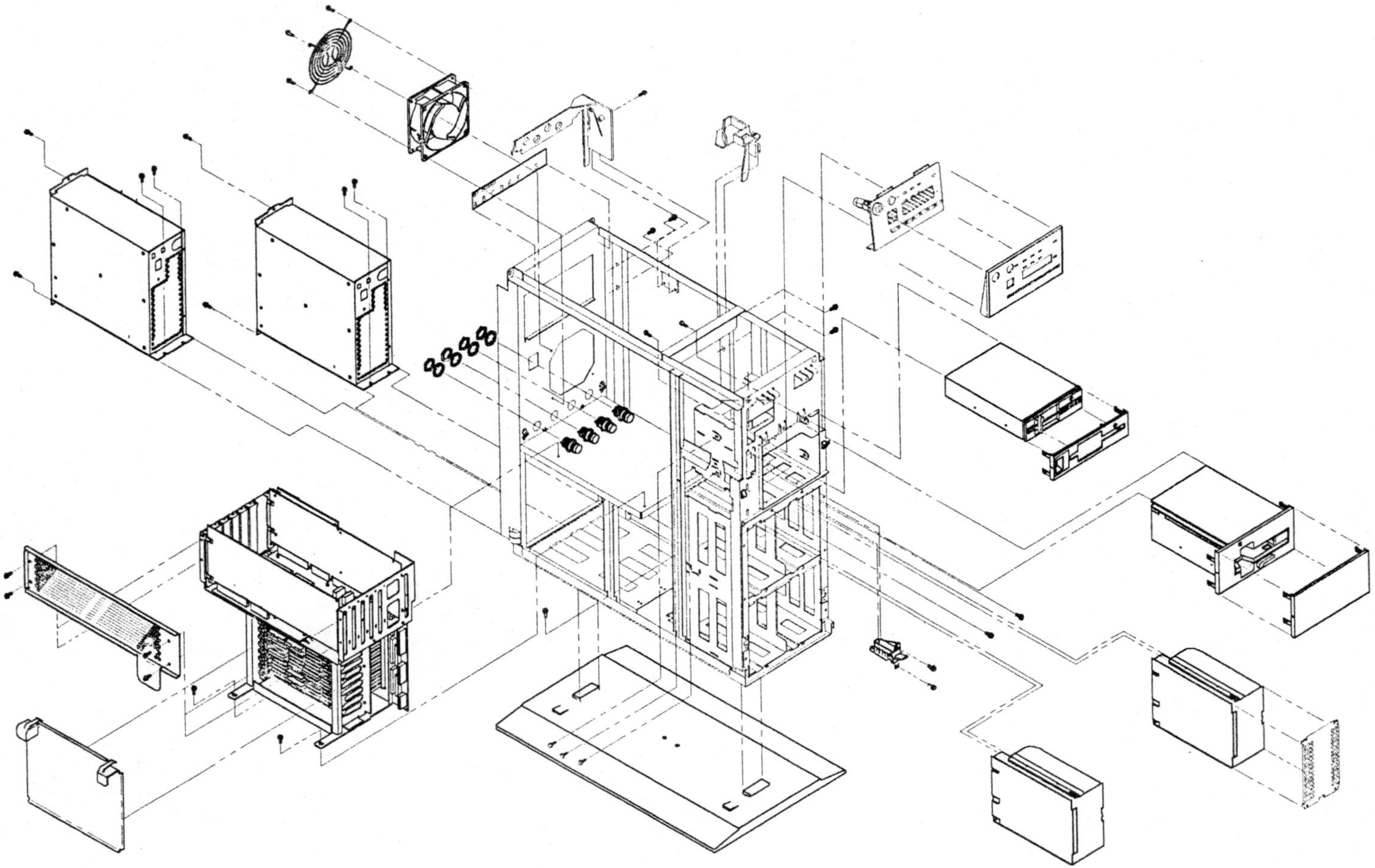

Figure 8.7 Mechanical hardware designed for the IBM System/36, Model 5363 by manufacturing engineers.

Utilizing the mechanical parts and process expertise of these manufacturing engineers was key to reducing the development cycle on this program. Manufacturability features were integrated on a real-time basis as design concepts were formulated. These "manufacturing" designers knew which components would typically be manufactured by IBM and which would be purchased. On purchased parts, they knew which suppliers had the best capabilities to be considered qualified manufacturers. They also knew how these qualified manufacturers designed and built their tooling. Since there are several acceptable methods to design tooling and manufacture parts, using this manufacturability knowledge as a designer avoided many design changes. Where functional requirements allowed, manufacturers' preferences were incorporated with only one design iteration.

Throughout the product design cycle, input was continually received from specialized testing and design reviews. The special testing included thermal, acoustical, fragility, and electromagnetic radiation tests. The design reviews identified design preferences from all plant functions previously mentioned, as well as marketing and service organizations. In many cases, the changes required had to be incorporated immediately to maintain schedules. There was not always time for a detailed design review by all of manufacturing prior to implementing the changes. The experience base of these "manufacturing" designers allowed incorporating changes on a timely basis in the most manufacturable method. The manufacturing process, tooling, and mechanical skills of these designers proved an excellent match for the mechanical design job.

Manufacturing expertise was a key ingredient in reducing the development/release cycle, reducing the cost, and optimizing the manufacturability on this product. Using the manufacturing skill base to do mechanical design was a very effective, efficient approach for IBM. Figure 8.8 identifies the cycle time reductions achieved on this project.

Conclusions

The boundaries between design, manufacturing, and service no longer exist. At IBM in Rochester, Minnesota, these functions are now part of an integrated product development process that ensures compatibility and reduces introduction time. Early Manufacturing Involvement at the concept level in the product design is key to optimizing function, manufacturability, cost, and quality. In the past, the value of involving manufacturing resources before the designs were well documented was not well understood or even considered. We have now clearly

Development cycle

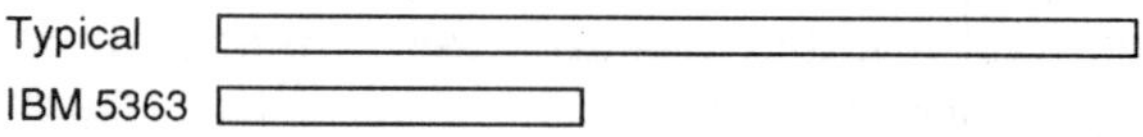

Typical ▭▭▭▭▭▭▭▭▭▭▭▭▭
IBM 5363 ▭▭▭▭

Manufacturing design team

- One-pass design

- Mfg. and procurement engineers responsible for mechanical design
    - Enhanced manufacturability
    - Utilized mechanical parts mfg. expertise for cost effective designs
    - Optimized cost, manufacturability, and function
    - Optimized EMI and engineering cross training

- Reduced development to manufacturing transition time with close working relationships

5363 results

Month 1	– Design start date
	– 75 parts are new design
Month 2	– Ordered first parts
Month 4	– First dev. level parts received
Month 5	– Built first systems
Month 6	– Released first designs
Month 7	– Completed mechanical parts release
Month 8,9	– Released required changes based on testing
Month 10	– Shipped early machines to customers
Month 11	– Met product announce date
Month 11	– Met first customer ship

Figure 8.8 Development cycle time reductions.

demonstrated that early involvement can assure that function and manufacturability are combined in a one-pass design.

Many lessons have been learned relative to EMI. A key item is that early involvement requires active participation in the design process; manufacturing should assume some of the design responsibility—gathering data, running experiments, and fabricating prototypes. Another important lesson is that prototypes should be built as soon as possible to find problems not easily identified with computer modeling. A third consideration is to begin functional testing with prototype parts as soon as possible. Many functional problems can be identified and solved prior to the availability of final production hardware. An additional lesson learned is that a small pilot line should be started as early as possible. Many manufacturing process problems can be solved utilizing early hardware. These early activities provide information that is critical to avoiding design changes and assuring over all cycle time reduction.

Realizing additional cycle time reductions and quality improvements in the future will necessitate increased integration not only of internal organizations but of suppliers and customers as well. The efficiencies achieved through integration of design and manufacturing can be enhanced by including suppliers and customers in the design process. Early and continuous involvement by manufacturing, marketing, suppliers, and customers will lead to products that meet and exceed customer expectations.

Chapter 9

A. B. Chance: Integration of the Design Process

H. Dennis Haubein

Introduction

A. B. Chance Company was founded in 1907 by Albert Bishop Chance, then co-owner of the telephone company in the city of Centralia, Missouri. Mr. Chance was an innovator and inventor. When he had problems with the anchors that he used to hold the guy wires for his telephone poles, he developed a device to solve the problem. In the past, he had used the traditional "deadman"—buried sections of logs in the ground, tied the guy wires to these logs, and tied the guy to the pole. In 1912, he invented a plate anchor—a metal plate that is buried in the ground at an angle to the guy wire; the guy wire is then fastened to the galvanized steel anchor rod. This was the beginning of the A. B. Chance Company's growth. Later came a series of innovations in power-installed anchors. These helical shaped anchors are rotated by truck-mounted installers and are screwed into the ground.

Through the years, the company has grown through acquisitions and product developments in product lines serving the electrical utility industry. In addition to anchors, product lines today include pole line hardware—the nuts, bolts, brackets, and other items required to mount products to the electrical utility pole; porcelain and polymer (plastic) insulators; hot line tools, used to repair and maintain electrical utility lines while they are energized; overhead and underground switches; and a variety of protective devices, such as fuses, reclosers, and other products that protect the lines, equipment, and the customer.

As the company has grown to 1,100 employees, it has become highly vertically integrated. Most of the parts and pieces required in the assemblies made today, are manufactured in the plants located in Centralia. This vertical integration includes forging, stamping of ferrous and nonferrous products, welding and fabrication of products, machining, a nonferrous foundry, and a variety of plastic processes used in making the hot line tool products. Some 15,000 catalog items are available today and many processes are required to make this product variety. Chance Company customers include electrical utilities and telecommunications companies. Chance markets products throughout the world, using its own sales force and a wide distribution network. Chance has always been forward-looking, adopting new processes to stay abreast of technology appropriate to the industry as it has become available. This has been particularly true in recent years with the advent of more and more computerized equipment in the manufacturing arena.

Company Design History

Over the years, many new products have been designed and moved into production in the Centralia facility. This design effort traditionally has been done by design engineering separate from manufacturing engineering. Manufacturing engineering then devised the processes to make the product, working closely with operations personnel and industrial engineers in the individual plants. Like many others, the Chance Company became more departmentalized over the years as it grew. This departmentalization isolated effort between design and manufacturing as well as the various production departments. As production departments were considered profit centers and responsible for their performance, the other nonproduction departments became entities into themselves with defined responsibilities. This traditional approach inadvertently raised walls between functions.

Reorganizing design

In the late 1970s, an effort was made to eliminate these walls and to make the design effort more of a company-wide effort. This was done through a team approach with PDAC (Product Development Advisory Committee) providing guidance for high-visibility projects. The PDAC group is the executive staff of the company, with design teams reporting progress at monthly meetings of this committee. Goals for product development are set by the team and approved by the committee. This effort used multidisciplinary teams working together to improve communication between the various groups and to improve the introduction of products into the manufacturing world. This was a very different way to organize the design process and it improved communication between all areas. Work was still done by individual team members with regular team meetings to ensure communication between the departments involved. There was still no guarantee of a close working relationship between the individuals, however. The closeness of the working relationship depended on the individuals on the team and how they interacted in their particular areas.

Computer-aided design (CAD)

Computerization at the Chance Company started in the 1960s with the computerization of the inventory control system. This effort continued to grow in other areas of the company with systems available for production reporting, purchasing, planning, order entry, accounting, and through evolution came the total business system as it is today. Early in the 1980s, a commitment was made by management to proceed into computerization of the design effort. This began with the purchase of CAD terminals to be used in design engineering. These terminals were used by full-time personnel, as well as temporary drafting personnel brought in from an outside firm. Three shifts were used to transfer the geometry from the "paper" drawing data base to the electronic data base. This effort continued for some two years and approximately 60 percent of the data base was placed on the computer system. At that time, the number of drafters was reduced and, rather than continuing to transfer geometry full time, it was decided to proceed with changing the remaining geometry over as revisions were made to drawings still in the "paper" data base. This procedure has been in effect since that time and today over 90 percent of these drawings are included in the data base. A very low percentage of revisions are not in this data base, indicating that the "coverage" of the active data base is essentially complete.

The key to the success of this program was that the drawing boards were moved out immediately and all work was done "on-the-tube." This contrasts with some companies who elect soft-conversion. That is, they do a mixture of manual and CAD for years and experience only a fraction of the benefits possible.

Training of CAD drafting personnel was conducted in-house. Two supervisors were trained by the software supplier and they then conducted all in-house training. Since the software was installed, a total of 53 employees have been trained. This included 9 temporaries as stated earlier.

Efficiency gains touted by CAD suppliers were very quickly realized. Figure 9.1 plots average graphics output per drafter and drafter headcount with respect to time. Examination of the curve shows steady improvement, with the pre-CAD output averaging around 10

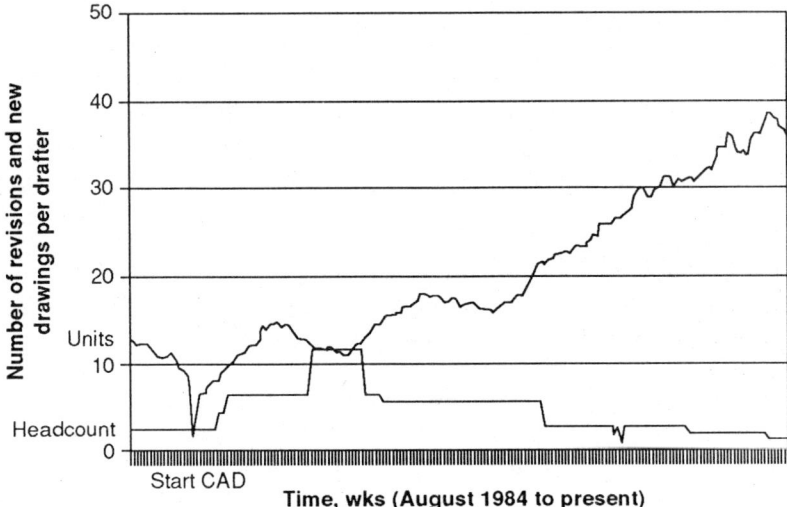

Figure 9.1 Average efficiency per drafter.

and the latest output averaging 35, or 3.5 times improvement. The headcount line shows the increase in total to minimize the time to get the geometry on the system. In addition, it shows that the number of drafters today is less than half the pre-CAD level. Examination of the output curve with respect to the manpower line shows two dips in output. One occurred immediately after the starting of CAD usage; the second, when the headcount was drastically increased. During training and again at the start of full time usage, drafters' efficiency dropped below pre-CAD, but in both instances efficiency recovered very quickly.

Figure 9.2 shows the use of CAD with respect to time. A 100 percent utilization is considered to be 8 hours per day and 5 days per week. Utilization of greater than 100 percent has been accomplished with flex shifts over a 12-hour day.

Computer-aided manufacturing (CAM)

With the computerized part-drawing data base becoming the majority of the total in 1985, it was decided to proceed with computerization in manufacturing. A CAM system was installed, in conjunction with the tool room, for the design of tooling, interfacing with the electronic data base in the design engineering department. This effort continues today with numerical controlled machine capability connected directly to the computerized system (DNC). In addition to this effort, Computer-Aided-Engineering tools, such as Dynamic Analysis, Finite Element Analysis, and Variation Simulation Analysis (VSA) used in

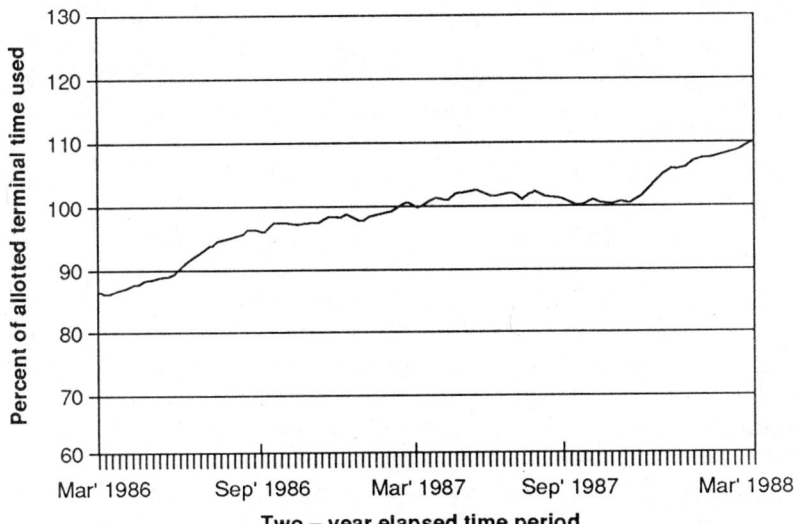

Figure 9.2 Average CAD use over 100-week period.

tolerancing, were also brought into the design engineering department. By the end of 1986, many tools were available for computerization of the design and tooling efforts.

Definite improvements were seen in the reduction of design time. This was particularly true when a family of products was being developed. Also, revisions could be made much faster. In-house employees trained include drafters, designers, and engineers. Use of the CAD system for design ensures compression of design time as the 2-D drawings can very quickly be generated from the designs.

Computer-integrated-manufacturing (CIM) and the design process

However, there was a recognition that the job still had not been finished. It was then decided that it was necessary to integrate all the computerized systems in the company through a CIM program. This effort started in 1987. A multiple disciplinary study was made of all the systems available in the company and what could be done to improve communication between them. From that study, guidelines were set and used to integrate all systems. In addition, it was recognized that just because the tools were available in design and manufacturing engineering did not mean that the job was finished in the improvement of the design effort. Availability of the tools does not ensure that the products that are designed and developed are, in fact, suitable for manufacture. As a result of this recognition, the total coordinated effort has four systems requiring integration. This includes the Computerized Inventory Control System, or CICS, which is the business system; the CAD system, including the computerized design and drafting system, which includes 3-D and Solids; the CAM system, which is the design tool in manufacturing engineering; and Design for Manufacture (DFM), the fourth quadrant or the fourth piece of the puzzle. Design For Manufacture is largely philosophical in nature rather than highly computerized. There are special techniques involved, but mostly involve communication and the use of common data in the various product and process design efforts. From adoption of the PDAC approach in the late 1970s, it was recognized that teamwork was very important in the design of products; the need for more attention to Design for Manufacture was also quite evident. However, some guidelines were needed to ensure simultaneous product and process design efforts.

The design guideline at A. B. Chance Company recognizes that *design is not the responsibility of design engineering alone*. Design of a new product is a company-wide initiative requiring input from many disciplines. To integrate this effort, teams representing these disciplines are directly involved in design. To define this team approach, a formal design guideline has been written. The guideline defines functional responsibilities, outlines the review process, provides checkpoints throughout the design program, and defines the steps in the design process.

The guideline is intended to serve as a tool for the product design team to ensure that all design aspects are considered from the initiation of concept to product availability. The use of the design team is necessary to ensure that all disciplines have input into the design and that manufacturability is considered throughout. It is recognized that not all design efforts require 100 percent adherence to each step outlined in the guideline or in the organizational phases. However, the chief engineer or the project leader is responsible for outlining a design program to ensure that the design criteria are met and the product is cost-effective.

Major product developments require that many outside inputs be received from concept through production of the new product. It is recognized, and all have read, that 60 to 80 percent of the product cost is set in the design, and that as much as 50 percent is set by the concept. With this recognition and the competitive nature of today's business climate, it is imperative that cost, reliability, and manufacturability be considered early in the design. This requires the up-front involvement of manufacturing, marketing, purchasing, and quality assurance in the design. This is accomplished through concurrent process design and continuous costing. All disciplines are involved as needed and included in the design review process.

Independent design effort in the past has resulted in numerous design changes after the release of the product to production. This is costly in both time and dollars. New CAD/CAM technology allows development of 3-D and Solid models from which costs can be estimated. In addition, these models can be analyzed using Computer-Aided Engineering (CAE) tools, and the 3-D models can be passed to manufacturing for use in the generation of tooling.

At certain predetermined points in the manufacturing of all complex products, tests or inspections of the product are made to determine compliance with the manufacturing requirements. Why? Because it has been found that problems detected at predetermined points in the process can be corrected more quickly and economically than problems that are detected after manufacture has started. This is true of all complex processes, including the engineering design process. *At predetermined points in the process, the design is checked for compliance with its requirements.* The objective is to ensure that optimum product design is achieved, considering all elements in the design that fulfill the customers' needs, and yields a satisfactory profit when price-competitive. The appropriate check for the design process is the formal design review. Therefore, the design process looks like that shown in Fig. 9.3.

Management of the Design Process

The foundation of any design program is the design specification. Therefore, it is imperative that a formal specification be prepared. The specification must be prepared prior to the initiation of the design process and a checklist is followed to make sure that all criteria are considered (see Fig. 9.3). It is important that the specification is available prior to team selection so that the team structure reflects the expertise implied by the specification.

Once a product design specification has been received from marketing, it is reviewed by the vice president of engineering and the chief engineer in the applicable product line. They then pick team members from various disciplines within the company, using the specification as a guide for needed expertise. Selection includes all team members. Even though all members may not be involved throughout the project on a day-to-day basis, it is imperative they be identified early in the project. In addition, this allows the members to know who their contacts are in support areas. It also has to be decided whether a team leader should be one of the members or not. The size and technical

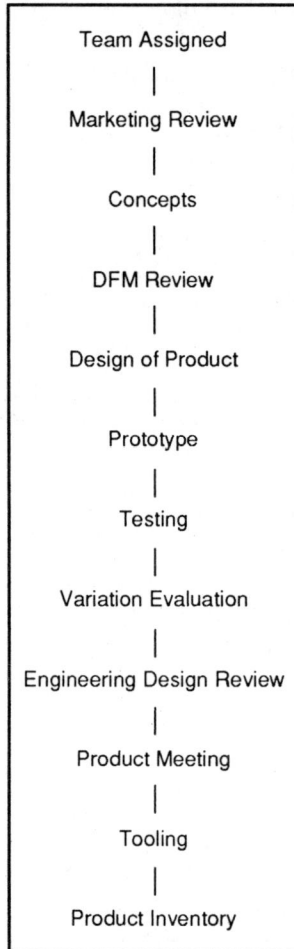

Figure 9.3 Design process.

complexity of the project may require an individual not involved on a day-to-day basis.

Design team structure

The structure of the design team changes as the project moves from concept through design to production. The core team members remain the same, but personnel are added or subtracted throughout the life of the project. Initially, the majority of the effort is in design engineering with manufacturing engineering participation and input from other areas. Manufacturing engineering participation promotes concurrent process design, costing, and manufacturability input. As the product moves closer to production, manufacturing engineering participation is increased. As the product is finalized, the complexion of the team changes, with the majority of the effort in manufacturing engineering and operations (see Chap. 3).

Most often the primary members on the design team include product design, manufacturing project engineering, and product marketing. Support members are then chosen from design analysis, product integrity, quality assurance, purchasing, production control, plant manufacturing engineering, shipping, and advertising. Detailed team responsibilities and the functional responsibilities of the team members are formalized.

The marketing review is a kick-off meeting for the design team. It is an opportunity for all team members to understand the scope of the project. The product marketing manager gives an overview of the market, introduces the key competitors, and covers the product specification. Design engineering is expected to respond to the specification and set the state of technology for the team. Their review should include any development already completed or contemplated. Concepts to be considered should be discussed, with goals set for the completion of the later reviews. One team member is selected as a team recorder or secretary so that minutes are kept of design reviews and that these minutes become part of the product design history and the project file.

Traditionally, design projects have started with drawing board layouts of product concepts, followed by prototyping of one of the concepts. Various design tests were then performed on the prototype and design modifications made until the product design met the specifications. After the release, design modifications were made per manufacturing request to allow for production.

With the advent of computer design tools, it is now possible to model several design concepts, cost them, and analyze them, using CAE software prior to prototyping. This reduces design iterations and shortens total time to production. Following the marketing review, the team moves on to this conceptual design. It is the responsibility of the design team to determine the steps to be followed in the conceptual design. Most major design programs start by modeling several concepts, using Solid modeling. These concepts should be costed and processes chosen for the lowest total production cost.

In the early stages of costing, many assumptions are made and rough costs are used. However, it must be kept in mind that costing at this point in design is primarily comparative in nature so that concepts can be evaluated. As design paths are chosen, these early costs will be refined and expanded until product release. At that time, complete routing sheets will exist and standard costs will be available.

It is the design team's responsibility to decide how far unique concepts should be carried. In some cases, more than one concept will be taken into the CAE evaluation, which includes cost and additional DFM analysis. The tools available include Product Costing, Process Evaluation, Finite Element Analysis, Dynamic Simulation Analysis, Variation Simulation Analysis (VSA), Statistical Problem Solving, Electrical Field Analysis, Design for Assembly, Failure Mode Effects Analysis, and Fault Tree Analysis. The depth of analysis required must be determined by the team.

DFM review

The primary purposes of the DFM review are:

1. To compare the design concepts with the initial requirements in the product design specification at a point where changes can be made most effectively
2. To compare the process design with requirements imposed by the design concepts
3. To obtain consensus on the design approach

Through periodic internal reviews, the team decides how many concepts to present at the DFM review to support the design direction selection. In some cases, they may want this selection to be made in the larger group at the review. The DFM review should be conducted after the basic product design is selected and preliminary layouts are completed, but before detailed product drawings are started.

At this review, it is the responsibility of the design team to present the results of the conceptual design effort and recommend the design approach. This recommendation should cover the cost comparisons and be supported by the CAE tools to show conformance to the original design specification. A DFM checklist can be used to identify the specific applications or areas in which each tool was used and should become a part of the project file.

Process design issues are as important as product design; concurrent engineering is a key factor for minimizing time from design to production availability. Process evaluation should include a study of capacity available if an existing process is being considered, and may include a review by production control.

The review process

The next steps—design of product, prototype, and testing—are combined for discussion purposes. These three steps in the design process should move rapidly, assuming success in the conceptual design phase. The Solid models previously developed can now be detailed as parts and assemblies with 2-D geometry and with 3-D wireframes generated, and used for computer numerical control (CNC) tooling of prototypes. As detailed drawings are made, the costs are constantly updated by a manufacturing engineering team member. This team member is responsible for generation of all costs and routings throughout the project.

As the design becomes set, additional participation by the industrial engineering groups in individual plants is mandatory to ensure that revisions are eliminated after release. It is the responsibility of the team leader, working with the vice president of engineering, to add manpower to the design team or increase participation of the members, as needed.

Prototype parts should be produced on production equipment as much as possible. This helps to make certain that part variations seen in later production are recognized as early as possible in the project. Total part variation must be considered because of:

1. design tests
2. costs
3. assemblage

Through all design iterations, cost updates are made, so that when the design is complete, the manufacturing documents are also complete. The design changes may also require updating of the DFM analysis previously done in the conceptual design.

A design team checklist is also made available to the team leader. It is the responsibility of the team leader to use this list as he or she sees it is needed. During any of the reviews, it can be requested by those reviewing and it is most useful as a tool for the team to verify that all points are covered. The checklist covers issues such as product safety, patentability, labeling, packaging, and so forth.

Before a product is submitted for the engineering design review, a decision must be made regarding the proper sequence of events to be followed once the decision is released for production. The design team must decide if a pilot lot of parts should be run to simulate total variation—taking into account production machine capability. This decision is dependent on the total number of prototypes produced and whether or not the expected production variation was represented by the prototypes. A sequence of tooling steps also must be defined.

After looking at the first three or four yearly requirements in the product forecasts, decisions can be made for interim tooling followed by hard tooling in later years. Once the sequence is known, the team must decide if the variation in the different processes can be represented in variation simulation, or if some pilot quantity run is necessary with variation measured to ensure that parameters are met.

Variation will be different depending upon the product, but is very important to be understood.

The engineering design review is held when the team members believe they have achieved the project's goals. The objective is an in-depth technical review with the necessary team participants and members of the engineering staff.

The product meeting is conducted when production drawings are complete, the process design is complete, and tooling can be ordered. The design and process can be compared with the individual requirements of the product design specification and steps should be taken to dispose of any unresolved items from earlier design reviews. All information related to the design should be documented and made available for backup, such as design specification, minutes from earlier design reviews, and all analyses. Particular emphasis should be placed on ensuring that the processes required by the design are capable of producing the design.

The responsibilities of the team are not completed until the product is tooled and the product is in inventory.

Problems, Results, and Expectations

As discussed earlier in this chapter, the training of individuals on CAD and the results of their use of the system have been documented. These results show better than a 3-to-1 efficiency improvement over the drawing board design methods previously used. With such efficiencies, come some management problems that must be addressed.

With the computer data base available, *drawing revision becomes so easy and fast that there is a tendency not to consolidate changes but rather to make changes as they become available.* For some time, the company has had change control boards whose responsibilities were to review all changes, regardless of where or by whom they were requested to make certain that they were, in fact, changes that were necessary and justifiable. In addition, the boards' effort was to make sure that changes were consolidated if the changes were not required to be made immediately. With the CAD reduction in revision time, these boards had a tendency to let all revisions move right on through the system and rubber stamp them. Therefore, it was necessary to retain some of the controls used in the past to ensure that these changes were not overwhelming the system. With the total system still being part electronic and part paper, these changes can very quickly overwhelm departments within the company. With the move to electronic release and 100 percent integration of this system, this will no longer be a problem. However, it is still felt that some control of changes is required.

Drafting time on new designs is reduced considerably with the use of Solids and 3-D in the design effort. Once a design has been conceptualized in 3-D or Solids and set, the 2-D drawings can very quickly be generated ready for release to the shop floor. Because of this, much of the drafting time previously required is eliminated.

Training is not necessarily learning

Training in the use of design tools, such as Finite Element Analysis or Dynamic Simulation Analysis, takes much more time and the results are slower to be recognized. As with any other new tool, successes are required in order to sell the use to the user. Initial use of these two design packages is primarily in the problem-solving mode. Problems are identified in the design of product during the prototype or testing stage or even as late as the production stage. Once these problems are identified, use of the design tools supports problem solving in that effort. In addition, production problems related to design can be quickly identified and used for this same problem solving. An example of such a production problem occurred in an extruded product. After production dies were built, the maximum electrical requirement could not be reached. Using the old cut and try method, the extrusion die was modified and electrical test rerun on the resulting product. This was done six times without the problem being solved. Use of Finite Element Analysis to solve the electrical stress problem was then attempted; the model quickly showed that the high electrical stress was located in a different area from that modified. Using the design from the Finite Element modeler, the problem was solved on the first modification of the extrusion die. This showed that design iterations can be reduced through the use of these tools. Another problem was identified in a switch mechanism which indicated that on closing there was arcing in the contacts. Using Dynamic Analysis modeling, a problem was identified that showed the electrical contacts closed but then bounced open. Once this was seen in the model, it was substantiated in the lab, using high-speed photography. Again, the problem was solved after the design was complete. These early successes substantiated the value of these tools.

Design philosophies: time and personnel

The challenge then became to move from the problem-solving mode into the design, so that problems are located and fixed in the design, as opposed to finding them after the design is complete. This is a philosophical difference. The tendency always has been to proceed quickly to a prototype, so that testing and design iterations could begin as soon as possible. People are now asked to do more analysis and conceptual work before any chips are cut or any prototypes are built. With the time and effort that is required to go into the Solids and Dynamic models, people need to have an understanding of what they are going to get from this time spent. This is a difficult thing for people accustomed to the philosophy of hurrying to prototyping. One of the primary things to learn is that there are going to be surprises in any modeling work that is done. These surprises in both the Solid models and Dynamic models show that the time required to develop them is time well spent. For example, Dynamic modeling teaches an understanding of mechanisms never appreciated just by building and testing mechanisms. Having the Dynamic model working in conjunction with Statistical Problem Solving, Variation Simulation Analysis and prototype models when available, creates a new understanding of mechanisms—an understanding never before seen in this company. There is an appreciation of what is happening in all situations in the model itself.

When moving into a team design, it is very difficult for product engineering or design people to let go and to learn to work closely with the outside disciplines. For so many years, they have been used to the idea of designing the products and passing this design on to other people for production. Now, they are asked to work with manufacturing people from concept to production. They are asked to turn over to these people more of the costing and evaluation work that they have wanted to be more closely involved with in the past. They still have a tendency to want to do it all, but as time goes by, and they see the benefits gained, the philosophy is changing considerably. They are now understanding that they can't do everything themselves. They understand the improvement in the results when input is received from all disciplines early in the design phase. Manufacturing engineering people, in particular, are working closer with design engineering on the entire design program.

Many people have asked why these manufacturing engineering people aren't moved to design engineering and become a part of the design effort. *The philosophy to date has been to have these people remain in manufacturing and be part of manufacturing*, working on manufacturing projects as well as design projects. This keeps them in the manufacturing organization and keeps them updated on changes in processes and improvements made in manufacturing, but makes them accessible to act as the liaison between design engineering people and the plant engineering people. The organization of manufacturing engineering is split between central groups and industrial engineering, which are located in the various production facilities. These people are responsible for the final effort in putting the product into production and working with the manufacturing supervisors in setting up these new products. The manufacturing engineering people in the projects, who are involved from day one, start phasing in the plant manufacturing personnel as soon as possible. They have contact people that they go to, so that there is someone they can get answers from in all cases. The manufacturing engineering people do the costing and the process evaluation, but they work very, very closely with time study, industrial engineering, quality assurance, and purchasing. All these people are involved in all the meetings, because it is necessary to make sure that they keep abreast of what is going on. It is critical that the support members be involved. They need to know what is happening throughout the project, not just in the review process.

Timing is critical. People need to feel the need for their input throughout. Teams need to be made up of decision makers, starting with the leader of the team and all the people involved throughout the team organization. They need to be available and they need to have the authority to make decisions. Also, it is essential that the leader of the team understands the various paperwork systems throughout the company and knows how new products are placed into production.

DFM and design philosophy

The Design for Manufacturing tools, such as Variation Simulation Analysis, Design for Assembly, and Statistical Problem Solving, have shown tremendous benefits when used in the design and conceptual stages. These tools let you evaluate changes and different concepts without building prototypes. As stated earlier, these tools support one another and work together to give total results required.

Figures 9.4 and 9.5 show how one product design evolved. The first vertical column is shown as 100 percent and represents the old design that the new product is replacing. The second column represents the first design that came from the design team; it shows a reduction in parts from 90 to 62 which is 69 percent of the original number of parts. So the people involved in the team intuitively knew that a simplified design and a better manufactured product could be achieved using less parts. The third column represents the product design evaluation following Boothroyd-Dewhurst Design for Assembly analysis (Boothroyd, 1987). It shows a reduction in parts to 47, as opposed to the original 90. This represents 52 percent of the number of parts without any sacrifice in function of the device. In Fig. 9.5, the first column shows the original design efficiency as 1. Using Boothroyd-Dewhurst Design for Assembly, the second column shows improvement in design efficiency of some 50 percent. The third column shows another improvement with a total

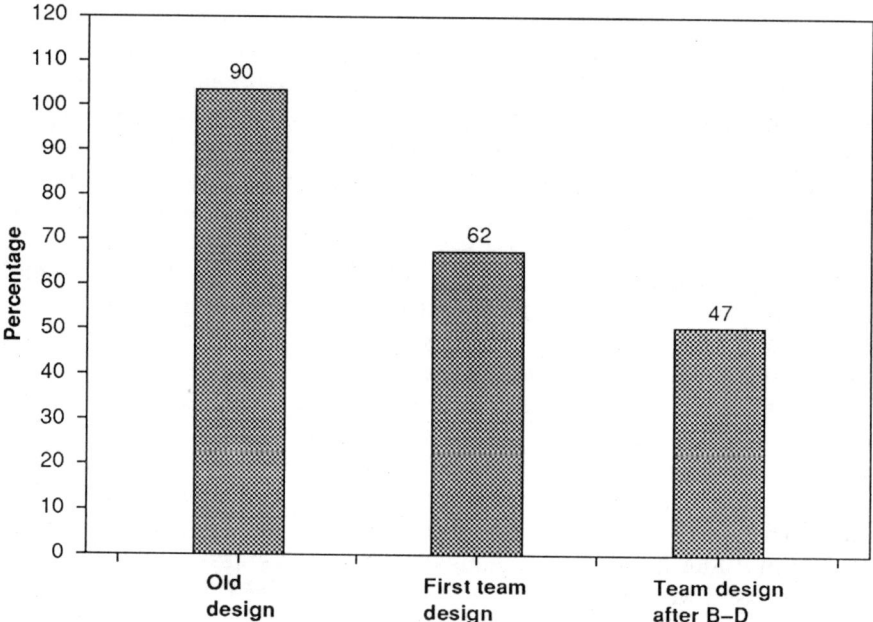

Figure 9.4 Improvement in reduction of parts.

of 75 percent improvement in design efficiency. These three columns correspond to the designs in Fig. 9.4. This new product replaces a design that requires considerable floor assembly and adjustment. These results are typical.

It becomes obvious that the results of the analyses support each other and work together to improve the overall efficiency in the project. This has been true in several instances, two examples of which will be discussed.

In the example previously used (Figs. 9.4 and 9.5) similar results were seen in the Variation Simulation Analysis* done on the same designs. One of the goals set in this project was to design an assembly that, after completion, would not require any adjustment for proper closing of the mechanism. To analyze this and determine if it could be done, VSA was used to determine whether statistically the parts could be used as produced in a stamping process and in a machining process with standard tolerances to produce an assembly that did not require adjustment. Obviously, the elimination of parts through the Design For Assembly analysis simplified the assembly and eliminated variation in the final assembly.

Examination of Fig. 9.6 shows results of the first run made with the

---

*Software for Variation Simulation Analysis used in this study is available from Applied Computer Solutions, Inc., 300 Maple Park Blvd., St. Clair Shores, MI, 48081.

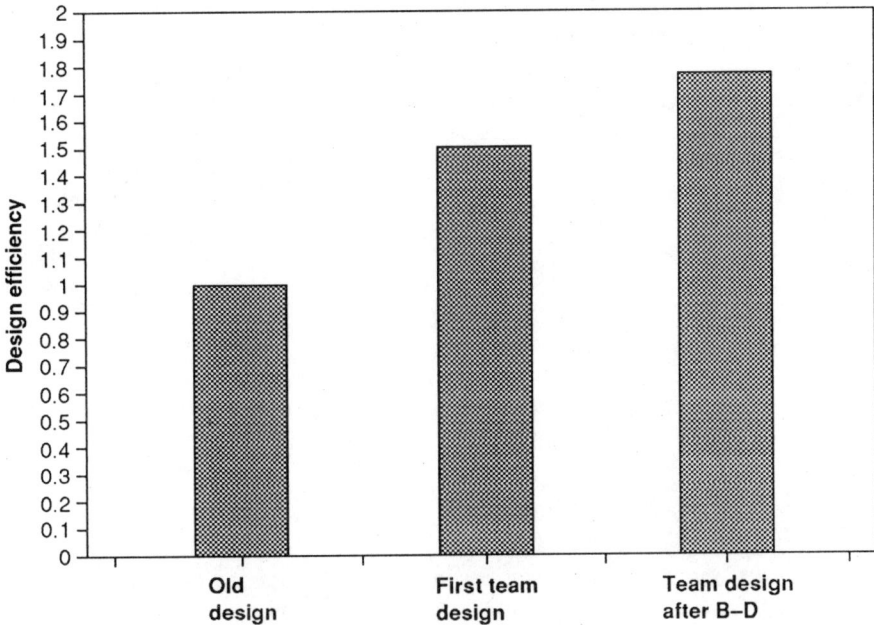

Figure 9.5 Improvement in assembly design efficiency.

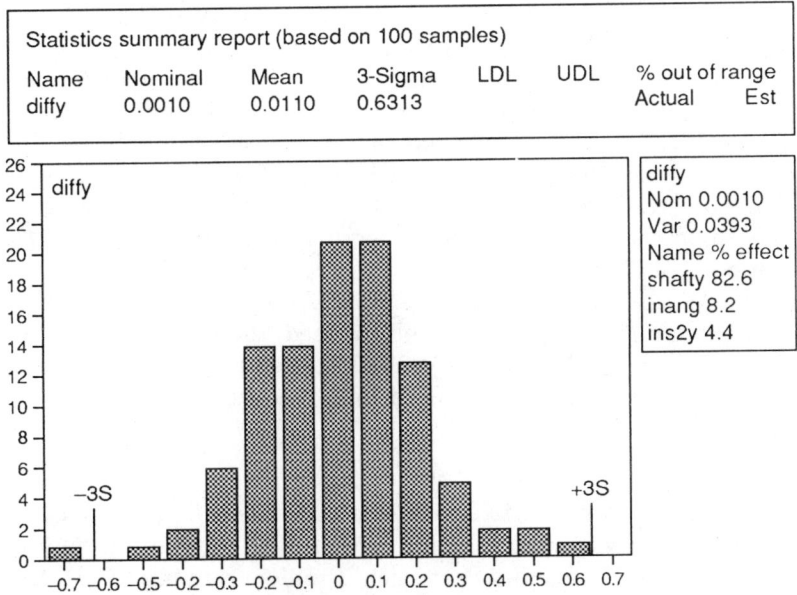

Figure 9.6 Initial variation simulation analysis.

Case Histories

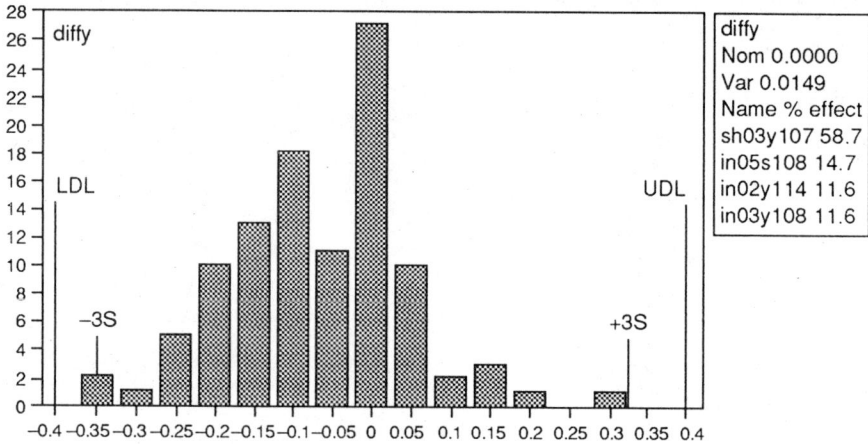

Figure 9.7 Variation simulation analysis after design changes.

design and initial tolerances. The major variation effect or the most significant variable is the shaft angle at 82.6 percent. The entire summary showed a 3-sigma value of 0.6313. After analysis of the highest variation and changes being made, the 3-sigma value in Fig. 9.7 was reduced to 0.3387. The shaft is still shown as the most significant variable but control of the shaft machining is possible because of the design changes and significantly reduces the effect of these variables.

The percent in Fig. 9.7, shown out of range because of the actual measurements, was 0 with an estimated range of less than 4 percent. With a process that is this capable, it is known that SPC control charts placed on that dimension will adequately control and keep the process in control. As stated earlier, these results are similar to the results seen in the Design For Assembly Analysis previously discussed. As the number of parts was eliminated, the ease of control of the various part dimensions was significantly reduced, and therefore, Variation Simulation Analysis showed improved results.

SPC and SPS

Since Statistical Process Control (SPC) has become the primary quality assurance tool used in A. B. Chance, significant training dollars have been invested in SPC and Statistical Problem Solving (SPS). Approximately 550 hourly people have received training in SPC and an additional 150 to 200 salaried employees have received training in SPC and SPS. Statistical Problem Solving uses design experiments to solve problems by determining the significant variables and then controlling these variables using SPC. Not only is SPS used in the manufacturing area to solve existing problems, it is also now used in the design area to determine variation and the significance of this variation in the design of the product. One example of such an analysis in which SPS was used in conjunction with another analysis tool, namely Dynamic Simulation Analysis, was recently seen. A mechanism product had been designed and prototyped and first article inspection samples had been provided to Design Engineering before the product went into production. Examination of various velocities and empirical Dynamic Analysis showed variation between the six first article inspection models.

A Dynamic Analysis model was built to examine what was happening in the mechanism. The dynamic model showed that two parts, which were intended to collide in the mechanism and therefore dampen velocity on closing, were actually stopping and bouncing back. Further examination using SPS determined why this was different in the six mechanisms inspected. A team of four people met and listed variables that could be causing the variation seen. These variables were then prioritized through a voting process to determine the top four. Measurements made on the six devices were then input into the SPS software to determine the most significant variables of the list of four previously screened. Analysis by the software substantiated what was thought to be the most significant variable. Modifications were then made to the mechanism levers and one lever that had been installed to do two functions was split into two levers, with an adjustment added for the functions to be separately made. Once this model was built and actual parts were made, additional measurements were made and this was then input to the software.

The results between the two design schemes are shown in Figs. 9.8 and 9.9. Figure 9.8 is the histogram and shows the results from the first run on the six mechanisms. Figure 9.9 shows the second set of data and the resulting change in operation of the mechanism due to the modification. The velocity variation in the first 25 measurements showed a calculated upper control limit in Fig. 9.8 of 4.126, which was well above the requirement of 2.95 feet per second. After modification, the second 25 measurements were well within the UCL of 2.95. In this case, the use of the Dynamic model and SPS working jointly along with the four experienced designers pinpointed the problem prior to its becoming a problem for the customer. This is just one more example of how problems can be found and solved prior to the manufacture of the product.

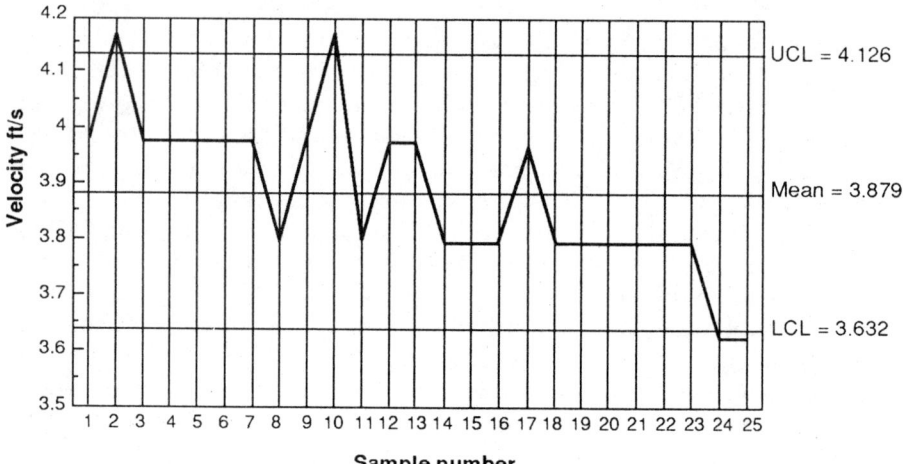

Figure 9.8 Results from the first run on the six mechanisms.

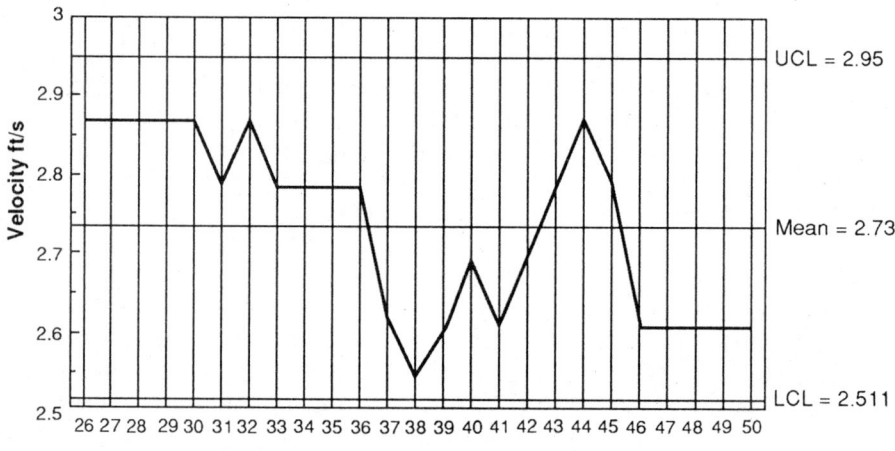

Figure 9.9 Results from the second run, made after mechanism modification.

Conclusion

Present tools available in this company have provided design insight and evaluation prior to "cutting chips" on prototypes to a degree never thought possible before.

Additional tools are being added in manufacturing engineering to enhance the costing and routing sheet generation on new products. An on-line computer system exists to set up released products and this system is being expanded to be available for evaluation during the process design. With additional changes allowing electronic release of drawings, only one data base will exist from design to the warehouse. With this integration, productivity, quality, and service to the customer will be further enhanced.

Computerization alone does not ensure that a company will be competitive, but, when integrated into a well-defined and managed design process, reductions in design time plus reductions in product revision after manufacturing release become a reality.

References

Boothroyd, G., and Dewhurst, P., *Product Design for Assembly,* Boothroyd-Dewhurst, Inc., Wakefield, RI, 1987.

Chapter

10

Northern Telecom: The Gate Procedure

Albert R. Wood and Paul D. Couglan

Introduction

In 1988, the president of Northern Telecom Limited (NT) described the company's vision of the year 2000. This vision demanded economical development and manufacture of telecommunications equipment products and systems "of the highest quality and reliability, on time, tailored to the varying needs of our customers." Effective management of the product development process was an issue in which the president would continue to be actively involved.

This chapter presents a case history of the application of a procedure, the Gate Procedure, introduced in NT in 1985 to facilitate simultaneous achievement of product manufacturability and tight schedule commitments. The Gate Procedure (or Gating) provided managers and product development teams in NT with an approach to aid and discipline them in making decisions during the new product development process. Experience with the application of Gating led to subsequent improvements in the procedure by individual divisions.

The theme of this chapter is that design for manufacture techniques and cross-functional development teams represent only two of three conditions necessary for achievement of product manufacturability and tight schedule commitments. The third is a disciplined management approach. Development teams need a step-by-step management procedure, or approach, to discipline their interactions and to take members through the trade-off decisions—among cost, features, and delivery—that crop up in virtually every new product development project. Such procedures should be dynamic in use, and harness the experience embodied in the development teams.

Northern Telecom

Northern Telecom is the largest manufacturer of telecommunications equipment in Canada and the second largest in North America. Their business consists of design, development, manufacture, marketing, sale and service of central office switching equipment, business communications systems, transmission equipment, telephones, cable and outside plant products, and other telecommunications products and services, mainly research and development. NT's competitors in an increasingly international market place for telecommunications equipment are Western Electric, Rolm and ITT of the United States, Ericsson of Sweden, Panasonic and Nippon Electric of Japan, Siemens of Germany, and Thomson of France.

The telecommunications sector has emerged in the last two decades as one of the fastest growing sectors in the world economy. By the year 2000, the global market for telecommunications equipment and associated services is expected to have grown from about $83 billion in 1988, to about $300 billion, a growth rate unmatched by any other industry. This growth is derived from a number of basic sources. Both traditional voice services and rapidly increasing requirements for data transmission are outgrowing the capacity of existing telecommunications networks to meet the competitive pressures of global business needs and the growing demands of society. At the same time, new technologies and new generations of equipment are both responding to and creating additional demands for new services.

In 1988, Northern Telecom employed 50,136 people worldwide, and operated 24 manufacturing plants in Canada, 13 in the United States, 2 in Malaysia, and 1 each in Ireland and France. Research and development was conducted in 24 of these facilities and by Bell-Northern Research Ltd. (Bell Canada), a subsidiary that operated R&D facilities in 10 locations including 4 in Canada, 5 in the United States, and 1 in the United Kingdom. In 1988, R&D expenditures were 13.1 percent of revenues of $5.40 billion, up from 11.9 percent of revenues of $4.91 billion in 1986. Selectively pursuing global markets, an estimated 80 percent of Northern Telecom's revenue was derived from products and systems introduced less than five years previously.

NT's operations were organized through five principal subsidiaries, including Northern Telecom Canada (NTC). NTC developed and manufactured many types of new products on an on-going basis. These products differed in terms of scale, complexity, value, flexibility, and backward compatibility with earlier generations of NTC products. Accordingly, NTC was organized into four operating groups. Each group came under the direction of a group vice-president, and comprised several operating divisions. The Subscriber Equipment Group had responsibility for design, manufacture, marketing, and distributing residential and business telephone equipment, digital private branch exchanges, and electronic key systems. The Group also repaired and refurbished telecommunications products. Five divisions belonged to the Subscriber Equipment Group, one of which was the Telecom Terminals Division. Many of the examples of Northern's approach to the management of new product development projects discussed in this chapter will be drawn from the experiences of a Canadian plant, the London, Ontario, plant in the Telecom Terminals Division.

Telecom Terminals Division

The Telecom Terminals Division manufactured residential, business, and public telephone sets, hands-free units, automatic dialers, and other related equipment. These products competed in markets in Canada, the United States, and throughout the world. While each of the telephone sets manufactured by the division varied in purpose and range of features, many materials and processes were common to all sets. Telephone sets went through a basic process flow, including injection molding of plastic parts, cord winding, and assembly of circuit boards, keypad, base, transmitter, handset, jacks, and cords.

Telecom Terminals Division—London plant

The London, Ontario, plant was built in 1959. By the late sixties, the plant was manufacturing most of its own telephone parts requirements and undertaking product development activities. The main product for the plant during these years was the rotary-dial 500 series Telephone Set—the ubiquitous "black" telephone familiar to all. From its introduction in the 1950s, this electromechanical telephone had been profitable because of level, high-volume production and continuous cost improvements. For most of the next quarter century, the 500 series formed the backbone of the product line at the London plant. Bell Canada owned over 10 million sets. With an average life of over twenty years, the 500 set was a piece of apparatus designed for longevity under severe environmental and operating conditions, and for ease of repair. Style and feature lines were added to the basic product line, and were produced in smaller quantities. By the late seventies, the original 500 set had been improved and modified to many "unique" applications and there was a wide range of models in production. In later years, however, the set became uneconomical to manufacture at the price the customers were willing to pay, and uncompetitive in the face of imports.

Teamwork and Techniques: A Response to New Technological Challenges

The sixties and seventies had been a period of relatively stable markets and technologies, not just for the London plant, but for NTC as a whole. Cycle times for new product development were longer than in the eighties. Normal practice was to develop and launch a new product in the market, and then, while in stable and routine operation, to improve yield and reduce manufacturing cost.

Following on this period of relative stability, the late 1970s and 1980s were, for NTC, periods of unprecedented growth and change. Major technological developments occurred in all three elements of the telecommunications system: distribution, transmission, and terminal services. Such developments were characterized by a combination of decreasing unit size and cost, increasing performance capability, and simplicity of operation.

The London, Ontario, plant was not immune to these changes. With the opening of the Canadian market to low-cost telephone sets manufactured offshore, competition for London moved from the traditional North American telephone manufacturers, which over the years had faced similar increases in material and labor costs, to Pacific Rim countries with wage levels less than 10 percent of London's. At the same time, the direct labor component of the 500 set represented 50 percent of the direct cost. By 1980, the London plant management recognized that, based on the 500 series sets and manufacturing systems, the plant was no longer competitive. The future survival and profitability of the plant depended on its ability to design, manufacture, and market in volume more cost effective, yet reliable, sets with high customer appeal, while meeting the stringent quality and systems requirements of the telephone companies. London had to simplify its product, automate the assembly, and increase the features. Harmony, a new touch-tone telephone set, was the first product in the fight back. Development commenced in late 1982.

Harmony was to have push-button dialing, be marketable worldwide, and be suitable for manufacture on flexible manufacturing systems. Further, Harmony manufacturing costs could be no more than 58 percent of the costs of the 500 series set. The in-house design and manufacturing engineering (ME) groups interpreted these objectives for Harmony in terms of minimizing the numbers of parts, fasteners, free leads, rotations during assembly, and materials types. They realized from the outset that, in order to meet these objectives, none of the existing 500 series parts or subassemblies could be utilized.

The designers and manufacturing engineers were asked to produce a design that would lend itself to automation. Most of the design and ME personnel had experience in component or product design, but had only been exposed marginally to design for automation. As part of their response to the challenge of the Harmony development, the London plant applied a set of technologies collectively referred to as programmable automation. The project team laid out the plant in robotic assembly cells. Assembly operations, which previously were completely manual, were transferred to fourteen robotic assembly cells. These cells were introduced in particular process areas including: plastic molding; assembly of transmitters, keypads, handsets, printed circuit packs (PCPs), and telephone bases; testing of PCPs; and telephone set packing.

Previous training in value engineering and on-going involvement in cost improvement of the 500 series sets assisted the designers and manufacturing engineers in their design of Harmony. The resulting design plan called for a total "vertical build" concept, where each successive product module (or major subassembly) would be placed on top of the preceding module, with no product elements inserted between horizontal layers. This change in philosophy made for more simple assembly, whether manual or automated. The improvements achieved through the Harmony design are illustrated by the comparison of the new Harmony set with the old 2500 set, summarized in Table 10.1.

Effective product and process development was not just a question of technology, however. The new technologies required a quantum leap in the precision of product and process specification, which was achieved through integration of the efforts of the design, ME, and marketing functions. These individual functions interacted continuously on the Harmony development in order to achieve integrated, informed, and balanced trade-offs among product cost, features, and delivery.

To design and develop every component and manufacturing process simultaneously with only limited engineering resources, ten cross-functional teams were formed under the direction of the division's director of technology. Appointment of the director as project manager underscored corporate management's view of the importance of the project. Each team contained a product designer, manufacturing engineer, and test engineer, plus quality and material control personnel. The project manager ensured that close liaison existed, not only among the cross-functional groups, but also with marketing, control, and manufacturing. The product development was given top priority

TABLE 10.1 Design for Manufacture: Harmony versus the 2500 Set

Key Item	2500 Set	Harmony
Total numbers of:		
Parts	380	120
Screws	24	7
Rivets	20	0
Free leads	24	2
Rotations during assembly	4	0
Metals	6	1
Manufacturing processes	10	3
Assembly operations	Manual	Robotic
Number of robotic cells	0	14
Relative manufacturing cost	100	38
Cost ratio: Labor:Material	50:50	20:80
Labor/set	24 minutes	6 minutes
Inspection	Batch	On-line
Final testing	100%	None
Manufacturing lead times	20 days	3 days

by all who worked on it, and, through teamwork and application of design for automation principles, the cost, delivery, and features objectives were met.

The London plant started production of Harmony in January 1984. Harmony was a success for the London plant and for NTC, both commercially and technologically. Through teamwork and techniques, Harmony drew the London plant back from the brink of closure. Yet, even while Harmony was being developed, the competitive environment had continued to change.

Compounding the new technological challenges, was a dramatic change in markets, caused by shifts in telecommunications regulations in North America, the United Kingdom, and Japan. Consumers and telephone companies required ever higher levels of quality and diversity that forced equipment manufacturers to upgrade tolerances and designs, eliminate defects, and manufacture a greater variety of products. Correspondingly, as product life cycles shortened, flexible, high-quality, cost competitive manufacturing was not enough: manufacturers had to accelerate the rate of new product introduction also. Companies were expected to replace whole product lines at short intervals. New, technologically advanced products were being introduced more quickly, with ever faster development cycle times, and with steeper production ramps.

In Canada, annual imports of telephones grew from 100,000 to two million sets between 1980 and 1984. The London plant management recognized that, no matter how successful, Harmony would never evolve as the 500 series set had, to support the plant for over a quarter century. To succeed under these conditions, they defined Harmony as the first of a new family of residential and business telephones that would be developed and launched on to the market at rates exceeding one per year. Other products using the same manufacturing system were introduced in late 1984 and in 1985. Frequent, short-cycle new product development projects were becoming a fundamental feature of the continuing operation of the London plant. Correspondingly, the management challenge facing the plant was not just to repeat the Harmony success, through teamwork and the application of design techniques, but to improve on it, and with increasing regularity.

Frequent, Short-Cycle Projects: The Gate Procedure Response

In NTC, the London, Ontario, plant was not unique in facing a need to manage the development of new products with ever-more ambitious schedules, specifications, and cost objectives, at increasingly rapid rates. The growing importance of product variety and technological sophistication as competitive factors in the eighties, and the increasing cost of maintaining that sophisticated variety, were pressuring NTC to improve all of their product designs for manufacturability through integrated product and process development. Further, development schedule overruns were resulting in greater loss of profit opportunities for NTC than cost overruns by either manufacturing or R&D during the development phase. Northern recognized the need to adapt their traditional approach to the management of new product development projects to the new competitive environment.

Until the early eighties, procedures for the management of new product development in NTC were characterized by many checkpoints and division-level reviews of progress. Northern emphasized both cost control and wise investment as essential for improved profitability. Correspondingly, as management decision-making emphasized finance, new product development was viewed largely in investment terms.

Northern gave each division, and each project team within it, responsibility for managing new product developments in their own way, as long as financial criteria were met and the product was competitive in the market. The development of Harmony had been managed successfully through appointment of a senior manager as project manager, cross-functional teamwork; application of design-for-automation and value analysis techniques, and an enthusiasm born of the need to survive. However, such an approach and its successful outcome would be difficult to repeat, especially in the context of frequent, short-cycle new product development projects. The elements of a sustainable management approach were present in the Harmony project (and in projects in other divisions); what Northern needed was some means of harnessing and formalizing the management experience embodied in the development team, and achieving the same result over shorter development intervals, without such concentrated input from senior management.

In general, companies that have tried to integrate design, manufacturing engineering, and marketing without changing the basic approach to the management of new product development, have encountered difficulties. There are barriers to the integration of these functions: location, background, education, budgeting practices, and performance measurement systems, to name a few. As noted earlier in Chap. 3, cross-functional teams are currently the most frequent way to cut through barriers for integration of design and manufacturing. While these teams include representatives of design and manufacturing engineering, in Northern they also included marketing and quality. However, the team is only the beginning.

Design for manufacture (DFM), as discussed in Chap. 4, is also necessary to integrate design, manufacturing engineering and marketing. DFM, a design discipline, consists of management tools and techniques, design principles and methodologies, and a philosophy of design integration and "global" optimization. Application of DFM aids a smooth transition from development to production. However, DFM and cross-functional teams represent only two of the three conditions necessary for integration of Design and Manufacturing. The third is a disciplined management approach.

Over a number of product development projects, a team learns to apply a design discipline in a way that reflects the company's design philosophy. Individual team members learn how to contribute to the philosophy through the creation of many informal communication networks. However, much of this learning is embodied in these individuals, and is lost to the team with their departure from the team. Further, despite this design discipline, teams still must confront trade-off decisions on every project—among cost, features, and delivery. Pressures and constraints associated with these decisions often dictate many aspects of product design, and make it impossible to follow best practices or achieve the best design. As such, teams need a step-by-step management procedure, or approach, to discipline their interactions and to take members through these trade-off decisions that crop up in virtually every new product development project. In a competitive environment characterized by increasingly frequent, short-cycle new product development projects, Northern recognized the need for such a procedure.

In 1985, NTC introduced a corporate-wide procedure, the "Gate Procedure," to manage the integration of new product development activities. The Gate Procedure was to provide guidance to all divisions on identification and achievement of an evolving system of technical and commercial objectives at each stage of the new product development process. While traditional financial controls remained in place, the new procedure was to facilitate divisions in expanding their views of quality in new product development to include absence of defects, ease of manufacture and operation, and timeliness of market availability.

From the start, planning for quality was to be an integral part of the product development process. Development teams were to catch and correct design deficiencies earlier. There was far more leverage in eliminating such defects early in the development process than later, when the product was in production or already in the market. Quality also meant reducing product components to their simplest form for operation and for manufacture. Unnecessary complexity added extra costs and reduced product and process reliability. Simplicity, on the other hand, led to improved manufacturability, market acceptance, and, ultimately, profitability.

Finally, quality meant timeliness of product availability to the customer. Throughout the new product development process, the Gate Procedure emphasized the achievement of delivery schedules. Frequent milestones and formal reviews helped impress on the cross-functional team members the urgency of their work. These reviews also allowed team members to foresee the impact of delays through the development process, and to control progress so that those dependent on deliverables late in the process would have adequate time to meet their deadlines.

The Gate Procedure was to provide a basis for improving the performance of succeeding new product development projects. By highlighting cost, time, and quality objectives for each project, and conducting formal reviews of progress at distinct stages, the procedure was designed to correct the difficulties of unsuccessful projects by placing an increased emphasis on cost avoidance. The intention was to create an environment in which divisions would experience continuous learning and adaptation. A summary of Gate Procedure features is outlined in Table 10.2.

Gates and Gate Reviews

The Gate Procedure divided the new product development cycle into four major stages: initiation, definition, development, and verification. The distinguishing feature of the procedure was a series of reviews at "gates," which occurred at the end of each development stage. Gates not only separated different types of activities, but also invest-

TABLE 10.2 Gate Procedure

Item	Gate 0 initiation	Gate 1 definition	Gate 2 development	Gate verification
Prime department	Marketing	Marketing	Design	Manufacturing
Product features activities	Product description	Commercial & technical specifications	Technical trial units	Field trial units
Product cost activities		Value analysis Yield target Subsystem cost target	Product cost Manufacturability New process qualification	Product cost
Product delivery activities	Program development plan Schedules & risks	Project management plan Marketing delivery commitment Design priorities Manufacturing/test plan	First ship dates Field trial plan	

ment foci with corresponding increases in the commitment of NT to the project. The corporate policy within NT required a minimum of three gates. More were optional.

A project could fail review at any gate. In this event, it could be cancelled, reworked, and again presented for review, or, if the reasons for failure were not perceived as major threats to the overall project, approval could be given to proceed to the next stage with existing inadequacies. NT recognized from the outset of its introduction of the Gate Procedure that the decision on whether or not to pass a project through a gate could not be rigid. Rather, passing was to be judged on a case-by-case basis, allowing factors, such as new technology risks and competitive pressures to get to market quickly, to influence the decision to proceed. However, the gate reviews would tell just how far a project was from its targets, and would indicate what needed to be done to put it back on track—before it became irreversibly late and expensive.

Gate reviews were carried out by panels of senior managers "appropriate to the level of importance of the project to the company." For some projects, the panels included the president and vice-presidents of NTC, and executive vice-presidents from NT. All functions involved in the product development were represented at the reviews, including marketing, design, manufacturing, quality, and customer service. The review panel focused on the product development schedule, product function, quality, cost, and manufacturability. They utilized check sheets to assess performance against detailed targets at each gate. The panel also sought to ensure that decisions made at earlier gates had been implemented, and that the project team had anticipated and dealt with down-stream problems that might occur in volume manufacturing and in the market.

"Prime" Responsibility

The Gate Procedure facilitated collaboration and teamwork among different functions. However, design and manufacturing were not the only functions whose efforts were to be integrated. Under the Gate Procedure, the marketing, manufacturing, R&D, quality, customer engineering, and installation departments were involved to varying degrees through the various stages of new product introduction, in either a "prime" or a "support" capacity.

A different department assumed "prime" responsibility for completion of each project phase on time and to specification: marketing and design at earlier phases, manufacturing, later. Transfer of "prime" responsibility took place at each gate, and involved certified completion of project responsibilities by each functional group involved. Under the procedure, a department would accept prime responsibility only when all preceding responsibilities had been discharged. In practice, departments accepted responsibility, even though specifications had not been achieved, and prior agreement was reached on subsequent product design changes of defined, but limited, impact on quality, cost, manufacturability, or schedule.

The Stages of New Product Development

The specific activities associated with each of the four phases of product development underlying the Gate Procedure corresponded generally to the four phases of design, described in Chap. 4 (see Fig. 4.2).

The initiation stage

The goal of the initiation stage was to identify a new business opportunity, which was reviewed at Gate 0, the formal beginning of the new product development process. Projects went through two phases in this stage: knowledge prebuild, and concept development. At the knowledge prebuild phase, planning groups, comprised of marketing and design representatives, matched market opportunities and available technology. From the many opportunities identified each year, perhaps a dozen with potential for developing into dominant designs were selected by corporate management for concept development. A full-time product manager and product designer then analyzed the market potential of the concept, and developed draft specifications and plans for project management and investment. At Gate 0, senior corporate and divisional management evaluated the new business opportunity as a cost effective and innovative technological solution that could be developed with sustainable margins on costs, revenues, and technology for a number of years.

The definition stage

Review of the product concept at Gate 1 marked the end of the next major stage in the new product development cycle. During this stage, the product concept was defined, and marketing set the context within which integration of design and manufacturing occurred. They developed a commercial specification, outlining the functional and aesthetic features of the product, its price range, and its market launch program, including required launch date. These items acted as commercial guideposts around which design and manufacturing engineering designed the product form, fit, and functions. The ultimate challenge for the development team was to remain within these guideposts, launching a price competitive product on time, while achieving the cost and technological objectives set internally for the product. Value analysis was coordinated by manufacturing engineering, in an effort to realize these objectives.

The review at Gate 1 was set up to answer the fundamental questions of any project: Do the designers know what the customer actually requires? Is there a clear understanding of the risks and opportunities in this proposal? Is there agreement on the strategy for commercial success? Is the commitment realistic? Is there unnecessary complexity? NTC saw Gate 1 as critical for any development project, in recognition of which, senior management, as a general rule, would not support any firm market commitments until this gate had been passed.

The development stage

The development stage ended at Gate 2. During this stage, the project team developed detailed specifications of what the product was to be, how it was to perform, what it was to look like, how it was to be manufactured, and how it was to be used. Here, product manufacturability was a major focus of the ME group from two perspectives. First, ME assessed the readiness of the design for later prototype production and testing. Second, ME assessed the product design for attributes that would both avoid production line stoppages, rework costs, and after-sales problems, and also increase safety, quality of workmanship, cost savings, and process compatibility. The focus on manufacturability supported a new corporate emphasis on design for cost avoidance through improved product development, in preference to cost improvement after product launch, with its associated engineering changes.

The verification stage

During the verification stage, which ended at Gate 3, the complete product was assembled by manufacturing, and tested under normal and worst-case conditions in typical customer application environments. The aim was to discover defects early and establish margins for failure modes, so that adjustments could be made as quickly as possible, before the sales and production ramp started. When the product met the commercial, manufacturability, quality, and performance criteria of the project team, the review panel at Gate 3 assessed the readiness to step-up production and ship to customers in volume. They evaluated all system test results, conformance to original specifications and plans, and issues of future product development and evolution including documentation, training, and product support. After Gate 3, the division released the product to the market.

The Gate Procedure in Operation

Both corporate management and project management reacted favorably to the use of Gating. Under Gating, top management were involved in new product development projects, functional responsibilities were formalized, and progress was reviewed at gates. Ultimately, better products were provided to the first customers. In a word, Gating worked.

Yet, the success of Gating was not achieved solely through increased formalization of the development tasks. In operation, the procedure was open and flexible enough to facilitate improvement of cross-functional integration, but in a way not explicitly mandated by Gating. Further, Gating was dynamic in use, evolving as it embodied the learning of the users in the various project teams. This evolution is clearly illustrated in the reaction of the London, Ontario, plant management to their experience with the development of the Elan telephone. These two features of Gating in operation, cross-functional integration and continuous improvement, will be discussed in turn.

Cross-functional integration and the gate procedure

Corporate management found Gating to be of significant benefit in keeping projects on track. Project team members gained a better understanding of project priorities as Gating encouraged process discipline, making risks and compromises visible for better control. Yet, in many respects, Gating was not essentially different from the management approach it replaced.

Under Gating, a project was reviewed at various stages of the development process, as discussed earlier. The Gate Procedure replaced, by major checkpoints and high-level reviews, the many checkpoints and lower level reviews that had characterized earlier approaches. Yet, project performance reviews were not new to NTC. The performance of the various groups involved in a project had always been reviewed in a number of settings. For example, each functional group reported on a regular basis to a director; the project manager presented progress reports regularly to all the function directors; the project manager chaired weekly project meetings, attended by all of the development team members. These meetings provided opportunities for cross-functional and top management interaction, which created awareness of issues, allowed evaluation of alternatives, and facilitated decision making and setting of directions.

The gate review differed from these meetings, however. Timing was event-driven, rather than time-driven; while the other interactions took place on a periodic basis, gate reviews took place only on successful completion of specified activities, such as prototype building, or completion of field trials. Even though the regular meetings reviewed progress, they could not stop a project for lack of progress; a gate review served that purpose. Yet, design reviews, and product sample reviews, held before the introduction of Gating were event-driven, and had power of veto.

Under Gating, a number of distinct groups were involved in a project, and included corporate, divisional, and functional management, the project manager, and the team members drawn from the various functions. Yet, these groups were involved in development projects even before the introduction of Gating. Gating formalized the responsibilities of each group and their interactions. Yet, functional responsibilities for product development were formalized before the introduction of Gating, albeit at a divisional level, and were not based on corporate-wide checklists. However, as discussed later, individual divisions adapted these checklists to reflect their experience with the use of the procedure.

In essence, the key difference between Gating and the earlier approach lay in the response of the project team members towards the procedure—they regarded gate reviews seriously enough to organize "dry-run" gate reviews in advance of the actual reviews. Gate reviews required consensus within cross-functional teams before a project could proceed. However, such consensus did not come about at the gate reviews.

In the London plant, for example, the project team members met in closed-session prior to each gate review. Only the team members attended. The objective was to examine in detail the preparedness of the team to go to the gate review. Each functional group presented, in draft form, the report of progress and problems that it had prepared for the gate review. The constraints, shortfalls, dependencies, and time requirements to satisfy the gate requirements were identified and discussed. By the end of the dry-run review, the team reached a consensus on the state of the project, the tasks to be completed in order to pass the gate review, the resources to be allocated, and the timetable to be followed.

The outcomes of these dry-runs varied from deferral of the gate review for a period of time, to proceeding according to plan. The dry-run reviews represented a new level of cross-functional integration in the product development process, and were a direct outcome of the existence of the Gate Procedure. In sum, Gating provoked an organizational response that served to achieve the objectives of the Procedure. Yet, there was no requirement for this dry-run review in the formal procedure.

Continuous improvement and the gate procedure

NT instituted the Gate Procedure to obtain impact beyond project management; it was to become a way of thinking within NT. They intended to create an environment of continuous learning and adaptation. Accordingly, the Gate Procedure never was intended to be rigid, recognizing that no individual set of management procedures could provide a unique and satisfactory solution to every problem.

In his earlier discussion on how and why DFM works, Stoll noted that as experience was gained with a particular design solution, "tricks of the trade" were learned that either made the solution work or improved its reliability or its performance. In similar fashion, the divisions in NTC learned to make the Gate Procedure work and to improve it. A summary of changes made in three divisions is outlined in Table 10.3.

Even though the Gate Procedure aimed specifically to achieve earlier consideration of manufacturing requirements, divisional experiences revealed that *the original steps underestimated how soon those considerations had to be addressed*. While the detail of each division's adjustments to the Gate Procedure reflected its specific competitive and operational circumstances, each division subsequently moved their assessment of product manufacturability to a stage earlier than originally specified. The Elan development project was a vehicle for this evolution of the Gate Procedure in the London, Ontario, plant. As a consequence of their experience with Elan, the London plant management added the requirement for an earlier volume production run, between Gates 1 and 2, on subsequent projects.

Elan was one of the first new product development projects managed in the London plant using the Gate Procedure. Development of Elan began in January 1985, and the product was introduced to the market in February 1987. Elan was configured differently from preexisting products within the plant, and it presented product and process design difficulties. The sleek and compact telephone featured a new electronic dial-in-handset. While automated assembly was the aim, Elan was not compatible with the automated assembly processes in place, a situation that worked against achievement of cost targets. As a result, product design and ME, through the project manager, requested that marketing either change the commercial specification or

TABLE 10.3 Gate Procedure Changes Initiated by Three Divisions

Change	Division A	Division B	Division C
Changed numbers of gates:	No	No	One extra gate between Gates 1 & 2.
Motivation for change:	Not applicable	Not applicable	Need to estimate yield and cost targets based on earlier pilot manufacture.
Basis for change:	Not applicable	Not applicable	Experience on earlier projects
Changed performance targets at gates:	New pilot volume production run before Gate 2.	New value analysis session before Gate 1.	Establishment of manufacturability targets at Gate 2.
Motivation for change:	Need for data, experience, and early opportunity to highlight manufacturability issues.	To evaluate product layout and definition, and to secure agreement of all functions on design stability after Gate 1.	Need for targets against which to assess the product at Gate 3.
Basis for change:	Experience on an earlier project	Experience on an earlier project	Experience on earlier projects
Changed transfer qualifications at gates:	No	Final assembly feasibility criteria.	Definition of prototype, eng. at sample, and preproduction unit.
Motivation for change:	Not applicable	Need for reduced cost, quality, safety, rework, & installation problems.	Need for agreed, measurable goals for manufacturability
Basis for change:	Not applicable	Experience on an earlier project (pre-Gating) with designs from Corporate R&D.	Differences in opinion within Division and with Corporate R&D.

the cost target, before reaching Gate 1. However, marketing, constrained to hold cost targets, modified the commercial specification, rather than accept a reduced margin. In December 1985, Elan passed through Gate 1.

Elan encountered further problems in advance of Gate 3. The field trial of units, prior to Gate 3, revealed a major product design defect, and review at Gate 3 was delayed until the defect was remedied. This process took an additional six months, during which time the complexity of the product and the pressure to meet launch date commitments increased. As a result, prior to the rescheduled Gate 3 review, ME were not able to meet the manufacturing cost targets. Further, they could not demonstrate that Elan's costs would decline from its high-initial level to the targeted level within the first year of volume manufacture without an investment program for cost reduction during that first year. Reaching the targeted cost level would require product and process redesign to improve utilization of labor and materials. Elan passed Gate 3 in December 1986.

If London management had delayed the product launch to allow ME to achieve targeted costs, development would have taken an estimated additional eight to ten months, during which time further capital investment and design resources would have been required. While delaying launch by this amount of time would have resulted in a more cost-effective product, Elan would have lost the market opportunity. Further, there still would have been uncertainty over product costs and yield; manufacturing and ME would still have much to learn about Elan through the operation of a volume manufacturing process.

As a consequence of the Elan experience, the London plant added an extra step between Gates 1 and 2: a pilot volume production run is now required for all products, using the intended process equipment, tooling and methods. This run "shakes down" a design before volume manufacturing start-up, enabling identification of manufacturability and product design problems that otherwise would not be found until volume production was under way. Material utilization is, correspondingly, greater and earlier than before, improving the opportunity to identify component-level design defects. This run was included at the expense of an increased development interval. However, ME successfully negotiated this change, in spite of pressure to decrease the development interval, because of its demonstrated need for information, experience, and highlighting of needed changes during the development process.

The Gate Procedure: Management Tool or Replacement?

The Elan case clearly illustrates that new product development remains an inherently uncertain process. This uncertainty arises from a number of sources, discussed earlier (Chap. 4) in terms of a feasible design region. Such uncertainty leads to particular difficulty in making estimates, for example, of the complexity of the product design, and its compatibility with preexisting processes in relation to cost, yield, and delivery targets. Yet, uncertainty can be managed through an emphasis on continuous improvement, and through a clear idea on the exact role of management procedures, such as the Gate Procedure, in the armory of the new product development manager.

However, a management approach, such as the Gate Procedure, is not meant to manage the development process. Rather, it should provide an opportunity for management to review the progress of the project and manage the trade-offs that always arise over the course of a project. A management approach will not screen out the mismatches in expectations, as represented by targets for product cost, launch, and features specification. A procedure is not like an expert system, that will do the thinking for management. In terms of the Gate Procedure, a project that passes a gate will not automatically meet subsequent specifications and other gate objectives.

A disciplined management approach, such as the Gate Procedure at Northern Telecom, is a powerful management tool to get to the heart of the project's health, to identify problems and to make the trade-offs that are necessary. When considered in relation to the objective of design-right-first time, the range of outcomes of these trade-offs constitutes a spectrum of alternative mixes of cost, features, and development cycle length. Yet, design-right-first time remains an objective to be worked towards. Such a capability evolves through continuous improvement, rather than management proclamation. In this sense, a management approach must be dynamic in use, evolving as it embodies the learning of the management users.

Conclusions

For even the most experienced managers, the development of a new or enhanced product presents challenges. If manufacturers are to take advantage of new market or technological opportunities, success in turning a promising idea into a finished product quickly, without sacrificing quality, demands excellent project management, and the integration of design, manufacturing engineering, and marketing. The market context of this functional integration demands key, but difficult, trade-offs among cost, features, and delivery. Design for manufacture techniques and cross-functional teams represent only two of the three conditions necessary for integration of Design, Manufacturing and Marketing. The third is a disciplined management approach.

Disciplined management procedures provide the basis for design, manufacturing engineering, and marketing to identify mismatches of cost and time priorities with their respective capabilities. Formal review of project progress, carried out by management from within and outside the project team, facilitates identification and agreement on tradeoffs among priorities. However, formal reviews are but one means of project review and functional integration. Project teams may also organize a set of informal reviews of project progress and priorities, without the involvement of top management, but with the express purpose of satisfying the requirements of the formal reviews by top management. Finally, management procedures provide a formal basis on which teams harness the experience of the various functions involved in projects, in order to make changes in their approach to the the management of the next development project.

Chapter 11

Implementing Simultaneous Engineering at Cadillac

Robert F. Jones

Introduction

Cadillac Automobile Company was founded in 1902. It is now a division of the General Motors Corporation. Cadillac facilities include ten sales zone offices across the United States, four manufacturing plants in Michigan, and administrative/engineering offices in the Detroit area. The entire organization involves the efforts of approximately 10,000 salaried and hourly Cadillac people.

In 1989 Cadillac was the number-one luxury car nameplate in America—a distinction maintained for 40 consecutive years. However, in the mid-1980s some considered Cadillac's position to be in jeopardy. In response to a pending fuel crisis, Cadillac had designed a new product lineup. The public's response to this lineup, however, was less positive than anticipated and Cadillac's market share began to decline. Cadillac needed to turn the situation around quickly. It needed to design and produce new vehicles that would recapture the Cadillac image and accomplish this in less than the normal time required for product development.

By 1989, Cadillac had done this and reversed the trend of declining market share. The turnaround occurred at a time when there were 37 luxury nameplates competing for a share of what was becoming the most lucrative and fastest-growing segment of the entire auto industry. An organizational transformation called simultaneous engineering (SE) provided the right environment for Cadillac to achieve this turnaround.

In the simultaneous engineering environment, new quality improvement processes that required cooperation across the organization

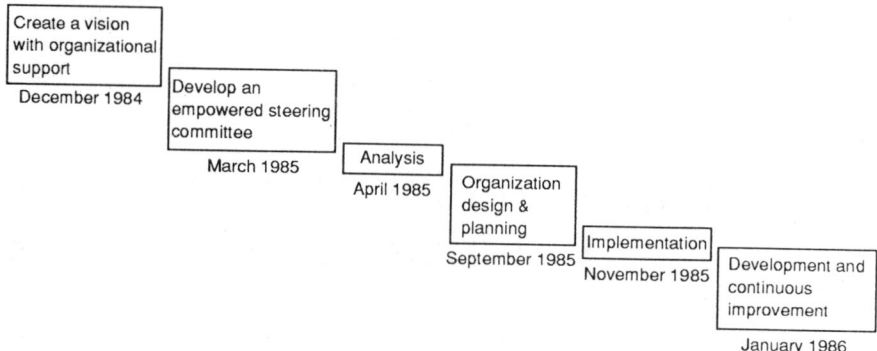

Figure 11.1 Implementation process.

wcrc more easily and effectively assimilated. *The redesign of the 1988 Eldorado, accomplished in an industry record 125 weeks, was just one of the results* of the newly developed simultaneous engineering environment at Cadillac. Another result was a high level of customer satisfaction resulting in improved market position.

The story of this transformation, the quality results, and a commitment to continuous improvement make Cadillac an excellent example to examine the successful implementation of simultaneous engineering.

This chapter is about the implementation and development of simultaneous engineering at Cadillac. Organization development issues addressed are leadership, strategy, structure, systems, processes, culture, and education and training. Each of these organization development issues are discussed during the chronological description of how Cadillac Car Division implemented simultaneous engineering (Fig. 11.1). The success of Cadillac's implementation of simultaneous engineering was due in part to the involvement of a large number of stakeholders in the organizational design and transformation decisions. With this important point in mind, the following discussions of organization design are not offered as a "cookbook" or fix for another company. For Cadillac, simultaneous engineering was right. The organizational leadership that was required and the change process that was followed provided the right catalyst. In addition, the organizational culture allowed the process to grow and succeed.

Create a Vision with Organizational Support

Staying on top in a mature market that is fiercely competitive presents serious challenges for organizational leadership. During 1984, Cadillac's leadership had decided to implement the concept of simultaneous engineering as one strategy to meet these challenges. This was the beginning of the change process. The organization's leadership began creating a vision. One style of leadership might have created a vision for the organization and then communicated it. However, the leadership style found at Cadillac called for the involvement of those who would be impacted in creating the vision. The result of this approach was a *shared* vision as well as the critical support needed for the successful change effort. A descriptive formula used by this style of leadership is Effective Decisions require both Accurate Information and Support (E.D. = A.I. + S.). In this case the people involved would eventually support what they had helped to envision.

In January of 1985, three workshops were designed and facilitated by a organizational development manager and co-chaired by executives from manufacturing engineering (process engineering) and vehicle engineering (product engineering). These meetings included not only the top engineering management but that of other staffs as well. At these workshops what was known about the concept of simultaneous engineering was shared. This management team then began to identify implementation issues. The output included a consensus to move forward, a proposed makeup for a Simultaneous Engineering Steering Committee (SESC), and a consensus to empower the SESC to act on behalf of the organization in further study and planning for implementing simultaneous engineering. Not only were these outputs significant, but this groups' involvement was the forerunner of the teamwork culture that would develop with simultaneous engineering.

Develop Steering Committee

An SESC was created consisting of what was called the "Junior Varsity." Each staff was represented by an executive one level below the company's executive staff. They met twice a month to determine their roles and responsibilities for designing and implementing simultaneous engineering. The newly formed SESC continued to plan for implementation. After a few months, they decided they needed someone working on their behalf full time. A crucial member was added to the steering committee. This member, called a "champion" was selected to act on behalf of the total organization. The champion, in turn, teamed up with the organization development manager who had conducted the earlier workshops and provided the full-time attention needed for implementing simultaneous engineering.

Some of the champion's responsibilities included: chair the meetings of the SESC, assist in the establishment and development of the rest of the simultaneous engineering teams (discussed later), coordinate management inputs from all the staffs as well as between teams, learn from other organizations' experiences with simultaneous engineering, and represent Cadillac in corporate simultaneous engineering assignments.

The role of the organization development manager included: provide input into the design and facilitation of the culture change required for effective teams, develop appropriate education and training for newly formed teams, and provide assistance to teams in learning processes for planning, assessing, and improving their team's effectiveness.

The champion and organization development manager together focused on the implementation process and provided the SESC with information, analysis, and proposals for continued learning and development of the simultaneous engineering process across the entire organization.

Analysis

In addition to listening to ideas from within the organization for implementation, it was also felt that Cadillac should learn from others' experiences with implementing simultaneous engineering. The organization development manager and champion began meeting with others in a similar capacity for other organizations. A network meeting took place monthly, then quarterly and provided many invaluable insights. These allowed Cadillac's future efforts to benefit from other organizations' previous learning.

The learning highlighted cultural issues, organization structure, physical location, education and training, as well as different business objectives as transformation issues for the simultaneous engineering teams. For example, some organizations had teams focused on future product development, while some were focused on current product improvement. Some teams formed in a matrix organization found opposition for their involvement from line management. In addition to the network attended, the organizational development manager and the champion participated in other organizations' simultaneous engineering education and training events to get ideas for Cadillac's future "kick-off" sessions.

Organization Design and Planning

In September 1985, the SESC and the executive staff participated in two simultaneous engineering vision and implementation strategy development workshops to assure alignment prior to establishing simultaneous engineering teams. One result was the shared vision developed for simultaneous engineering as follows:

Simultaneous engineering is a process in which appropriate disciplines are committed to work interactively to conceive, approve, develop, and implement product programs that meet pre-determined Cadillac objectives.

A further development was the pyramid structure (Fig. 11.2). Cadillac has adopted the pyramid as the symbol of Simultaneous Engineering. At the base or foundation, is the Cadillac executive staff who support and nurture the process with the ultimate objective of satisfying our customers—at the top of the pyramid.

The role of top management in the simultaneous engineering environment is to:

- Sanction the simultaneous engineering process
- Set simultaneous engineering policy and direction
- Provide the environment in which simultaneous engineering can flourish

Any time an organization sets out to make a significant change in the way it does business, it is going to take a great deal of time and

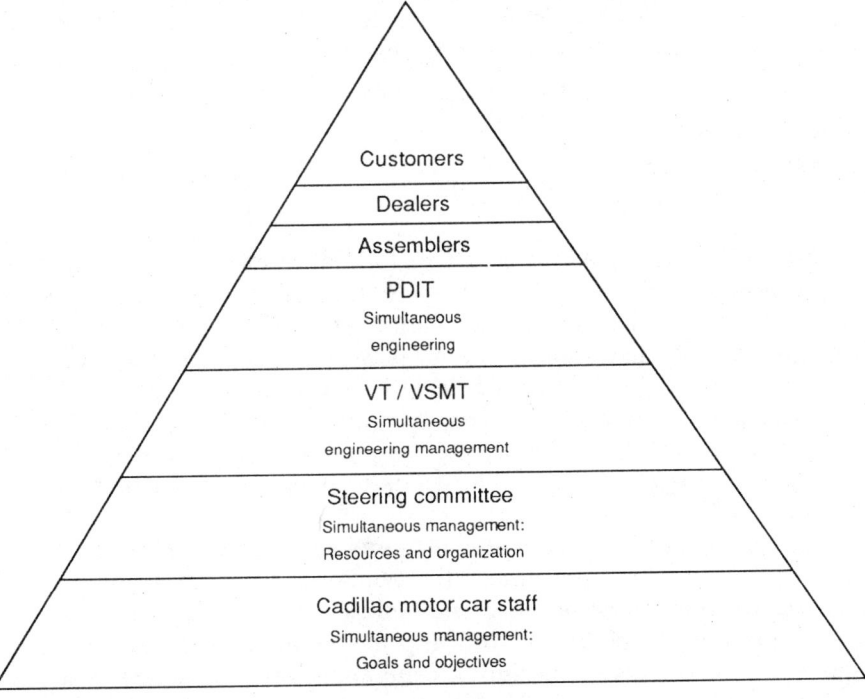

Figure 11.2 Cadillac simultaneous engineering pyramid.

education for all employees to make it work. But, without top management's leadership, support, patience, and commitment nothing will be accomplished. Next on the pyramid is the steering committee whose job is to:

- Plan and implement simultaneous engineering policy and direction
- Allocate the necessary resources
- Serve as liaison to communicate the process to the total organization
- Monitor and lead the process

Next are vehicle teams that are responsible for managing all steps of product development in their vehicle program. Each vehicle team comprises members representing all staffs of the organization. The roles of the vehicle team are to:

- Develop the vehicle strategy including defining the target market and specific demographics. This vehicle strategy must be consistent with the overall divisional strategy.
- Establish the overall vehicle goals required to meet this strategy.
- Manage the vehicle content. Provide complete, consistent, stable, and timely program definition for each vehicle.
- Assure the needs and expectations of the customers are met or exceeded.
- Manage the continuous improvement of the vehicle's quality, reliability, durability, and performance.

As Cadillac developed the structure for simultaneous engineering, the car was sectioned into specific vehicle systems and created six corresponding vehicle system management teams. These were the exterior component/body mechanical, chassis/powertrain application, seats and interior trim, electric/electronic, body-in-white, instrument panel/heating, and air-conditioning systems. The role of each one of these vehicle system management teams was to manage their vehicle system in order to optimize the business decisions that are made in that area of the vehicle.

The vehicle system management teams and the vehicle teams are in the same layer of the pyramid. This symbolized their partnership and interdependence to accomplish the task.

The product development and improvement teams (PDITs) are responsible for the actual design of components that are part of the six vehicle systems. Each PDIT has varying core memberships, depending on its purpose, but can draw members from any area of the organization and suppliers. One hundred percent of the vehicle is covered by these simultaneous engineering teams.

In some companies the simultaneous engineering approach calls for product development teams (PDTs). These teams include process and product engineers in the development phase of products, then disband when the particular product goes into production. Unlike these PDTs, Cadillac's *PDITs have cradle to grave responsibility* for the productions and continuous quality improvement of that component or part. Cadillac PDITs focus on all business aspects of their assigned portion of the vehicle: quality, cost, timing, technology, reliability, and profitability. It is as if they are running their own business. Cadillac eventually created 66 PDITs with an average of eight team members.

The structure of the pyramid is similar to a matrix organization structure although Cadillac has formally maintained its centralized functional structure. Each team member still reports to a staff area and has other assignments as well. With the exception of the vehicle teams, all other simultaneous engineering teams elect their own chairpersons and do not have a manager as in a typical matrix structure. The teams receive expectations and leadership from the next team down in the pyramid.

Each of the vehicle systems management teams is responsible for business decisions concerning its systems, as well as determining what vehicle subsystems require the formation of PDITs. Each PDIT, in turn, has similar business decision-making responsibilities at a component or subsystem level.

The vision was developed and the structure was determined. Roles and responsibilities were defined and the strategy for simultaneous engineering was ready for the next stage of implementation. The new expectations of team members would require them to learn about other parts of the business. In addition, most team members were familiar with planning and decision-making in the context of their individual staff, but not with cross-staff teams. Normally this type of decision-making is not experienced in a centralized organization except at the executive staff level. The need to develop consensus decision-making skills and teamwork was acknowledged. A great need existed to provide education and training.

Implementation

Change takes time and education. In November of 1985, an organization event was held to communicate the plan. It was considered important to communicate the design for simultaneous engineering to those who had originally met in January as a follow-up since they had empowered the steering committee. It was also considered important to communicate to significant others who would eventually be called upon to staff the simultaneous engineering teams. The meeting was designed to be interactive. All questions were documented and a response was given either by the panel of SESC members, executive staff, or included in forthcoming documentation of the worksession.

At the November meeting, the idea of forming Vehicle System Management Teams (VSMT) based on various sections of the car was shared. At this time each staff representative on the SESC began to consider who should be assigned from their staff to form these new simultaneous engineering teams. Although each staff retained the authority for their own selections, they received input from others on the SESC. Early in 1986, four-day kick-off workshops were held for the formation of VSMT. The design for these workshops included five sections:

1. Background Information
2. Cultural Change
3. Business and Systems Information
4. Planning
5. *Esprit de Corps*

The executive staff and SESC demonstrated leadership and support by participating first in the workshop in January, followed by VSMTs in February. The final day of the VSMT workshops included highly creative nontraditional presentations that were attended by both SESC and the executive staffs. The final day presentations were an organizational event. There were celebrations and the demonstrated enthusiasm further nurtured the evolving teamwork culture.

The newly formed VSMTs began identifying appropriate product development and improvement teams for their system. Appropriate members were notified of their selection and in April, PDIT kick-off workshops began. They were similar to those for VSMTs but included more emphasis on problem solving techniques. They, unlike the four-day VSMT kick off, were delivered in three phases:

Phase 1: Simultaneous engineering, business, and systems information

Phase 2: Team building and planning

Phase 3: Problem solving (applied to product quality)

Development and Continuous Improvement

As each simultaneous engineering team completed the kick-off education and training session, it left with a plan. These plans included goals and action steps to achieve business objectives for their team's specific responsibility. Vehicle teams focused on a total vehicle (e.g., Eldorado). Vehicle system management teams focused on their vehicle system (e.g., Body-in-White). Product development and improvement teams focused on their product (e.g., deck lid).

As the simultaneous engineering process developed, some formal organization structure alignment was necessary. In addition, some changes were needed for systems and processes. Some examples of formal organization structure alignment are as follows. Vehicle engineering combined all sheet metal design into one department to better align with manufacturing. Materials management and customer satisfaction reorganized to have individuals' functional responsibilities align with SE team assignments. Next are some examples of systems changes. Business information that had been tracked by plant or staff was needed by the teams. The information system was adjusted to include tracking and reporting information by vehicles, VSMTs, and PDITs. Another system that was aligned to support simultaneous engineering was the reward system. A reward and recognition system was developed to recognize the efforts of teams and encourage teamwork.

Some processes also required realignment. Examples are a quality improvement process, a cost reduction process, a product development process, and an assessment and planning process. Each of these and others were developed or modified to engage the simultaneous engineering teams as an integral and value added element of the business. In the spirit of continuous improvement, many of these processes were modeled for initial understanding and then subsequently improved.

It's important to note that the *implementation alone is not enough*. Organizations are too complex for the simultaneous engineering change effort to be viewed only as a linear model. At Cadillac, developing simultaneous engineering is a dynamic process; one which does involve vision, planning, and execution, but does not stop there. Yearly, the SESC, as well as the other simultaneous engineering teams, assess the success and opportunity for improvement to the structures, processes, and systems, and incorporate what is learned into plans for continued improvement. This approach requires a culture supportive of honest and open communication of information with an urgency to continue to improve the quality of its products, processes, and people.

Learning

Cadillac, through simultaneous engineering, institutionalized teamwork. A formalized simultaneous engineering structure was essential. Education and training was required, and management commitment and involvement was instrumental in effecting a cultural change—one that fostered teamwork, communication, and group decision-making. A key to Cadillac's success was the teamwork and communication that took place among the many different staffs in the various stages of the development process. There also was active participation of stakeholders at all levels of the organization (that is, Cadillac Motor Car staff, SESC, VSMT, VT, PDIT, and so forth) in the decision-making process.

The implementation of simultaneous engineering was well served by involvement of a critical mass of supporters large enough to sustain a change effort and overcome the "body/culture at rest tends to stay at rest" analogy. The three meetings in the December 1984–February 1985 time frame helped create a shared vision by a critical mass of the organization's leadership. In addition, the kick-off sessions held January–August 1986 helped create a critical mass deep in the organization focused on the planning and execution of the simultaneous engineering process. An empowered steering committee that demonstrated a trusting and participative leadership style and learning from other organizations' simultaneous engineering experiences was also key in planning for a successful implementation.

Simultaneous engineering is now an important element in aligning the organization's business objectives. Through simultaneous engineering, there is increased focus on people, process, and systems, as well as on the product. All levels are empowered and take personal responsibility for leadership for their part of the business.

The positive effects of simultaneous engineering are being experienced at all levels of Cadillac. People now wear two hats: their traditional functional hat and an SE team hat which, when coordinated by Cadillac's business plans, encourages everyone involved to contribute as if they were running their own business. This involvement has had a tremendous impact on Cadillac's ability to improve its products and services. Open, two-way communication and continuing education is working. People share knowledge, giving each other information that will help everyone do their job more successfully.

Breaking down barriers, pooling resources, getting input from those affected by decisions—that's what simultaneous engineering is all about. What started out as a new way of engineering a vehicle developed into a whole new culture at Cadillac. Simultaneous engineering today is a primary force in Cadillac's position as America's luxury car leader—a distinction they have maintained for 40 consecutive years. In the long term, Cadillac believes simultaneous engineering will help to maintain the focus on the achievement of its mission—to engineer, produce, and market the world's finest automobiles, and to continue as America's luxury car leader.

Part 3

Baseline for the Future

Chapter 12

Integrated Design Management

Introduction

Design has a life of its own. The essence of this book has been to introduce a new perspective on design so that this new design-life will be better understood. Our theme is quite simple: at the heart of this new approach toward managing the design process is a philosophical *shift* from doing one thing at a time well to doing many things well all the time. The implication of this shift is that design and manufacturing will now work as a closely coordinated effort, but they will still maintain their autonomy. Furthermore, other functions will orbit and take direction from this fundamental philosophical shift at the core of the firm. This implies that there is an optimum level of integration between all the functions of a modern organization, and that each interface has its unique characteristics.

The purpose of this final chapter is to take our experience, our data from surveys and the literature, and the cases reported in the book, and synthesize them into an understanding of how to plan and execute this new philosophy. These new principles include an appreciation for the time that is required to learn new design approaches, a sensitivity to the role of both ad-hoc and permanent changes in organizational structure, a vision of the design office of the future, and a discernment of the relationship that evolves between design and the business enterprise. The end result of these new directions is, no less than, a passion for design in manufacturing.

New Principles

In this section, we detail these new design management principles. For each of these new ideas, we summarize the insight upon which they are based, we explore their implications, we illustrate them with examples, and finally we make recommendations about how these new principles can be used.

Reflection on time to design

We are concerned about the wide-spread advocacy of techniques like simultaneous engineering as a means for reducing the amount of time it takes to go from an idea to a realized product. Our experience—the case histories of Northern Telecom, Cadillac, A. B. Chance, IBM, and others—strongly suggests that this advocacy is misplaced when these techniques are being applied for the first time in a firm.

It takes time to learn these new techniques. What is more, this learning is predicated on the assumption that the organization has already made a philosophical shift to a more complex view of the forces that influence excellence in design. We are not opposed to simultaneous engineering or any other approach that promotes design-process integration. We strongly advocate these techniques in Chaps. 3 and 4. However, our advocacy assumes that it is applied to the right product at the right stage of the life-cycle and that, at a very minimum, core team members have made this shift in philosophy. This shift is tantamount to a religious conversion process. Successful consummation of product-process integration is produced by people who have made this conversion. But, long periods of training and development and significant attitude shift are required.

How long does it take to go through this conversion experience? In our experience, it usually takes a year to eighteen months minimum, and a more representative estimate might be two years or more. What impressed us about the Black and Decker case, the Northern Telecom case, and others, is that the philosophical shift resulted from the widespread realization that continuing on the current course of action would result in irreversible loss of market share and noncompetitiveness. Companies that feel secure will not change quickly.

You can't start this regeneration with a technique and you can't afford to put off starting. Product-process integration is learned by doing and it takes time. There are many starting points for this philosophical shift. For example, a quality initiative or a competitive jolt, can often accomplish this. The important thing to remember is that the goal has to change.

Therefore, we recommend the following:

- Don't begin simultaneous engineering until the philosophical shift has occurred.
- Don't use simultaneous engineering for the first time on products

that have to be redesigned or launched in short periods, that is, in less than two years.
- Don't rely on simultaneous engineering alone to carry your design-manufacturing integration effort.
- Do think of design-manufacturing core team activities as part of a broader technology revitalization horizon in the firm.

Organizational structure

Organizations in transition require ad-hoc structures. No one in an organization, even those people with incisive vision for the future, can see all the details or ramifications of the philosophical shift. Our experience has shown that once this shift has occurred, things that weren't possible before become viable alternatives. What is more, when people pick a new permanent organizational structure, they tend to use earlier generations of the firm as a model. For example, we have observed many companies that have reinstalled manufacturing engineering reporting to engineering. Typically, this does not work as well as a temporary team structure to integrate design and manufacturing, especially when there is also a careful selection of the team leader. Not only does a more innovative structure result, but it is more fluid by definition and it can evolve into the future.

The core team is augmented from time-to-time by the infusion of temporary members from various functions of the firm. These membership patterns plotted over the duration of program efforts have been discussed in Chaps. 3 and 7–10. Methods for auditing this process are in Chap. 3. But the point is that the core team always directs the effort and the other functions support it. Therefore, we recommend the following:

- Tailor the structure to the task: small projects require minor shifts in task assignments and travel, large programs require significant task force management.
- Accounting and MIS ought to be part of the program effort but should not take the lead in directing the core team in the selection of major technologies like CAD/CAM and networking software.
- By the same token, the core team has to participate in organization-wide efforts to establish standards.
- The more outsourcing required for the product, the earlier that purchasing and suppliers need to be involved (see Chap. 6).
- When investments in new tooling, equipment, or software are planned, the suppliers of this technology need to be involved with

the core team—after the goals of the program have been clearly stated and general requirements have been determined.

Disciplined anticipation

We are impressed with the A. B. Chance approach to a disciplined framework for the design process. In Chap. 4, we call this design process improvement. *We are struck by the absence of discipline in the design process as practiced in American manufacturing*. The Gate Procedure at Northern Telecom and the project management philosophy used at G.E. are good examples of a disciplined approach. In our experience, these are exceptions.

Once a company has begun on the road to the philosophical conversion process, the way its members successfully anticipate the future is to adopt some type of disciplined design process that makes sense for the company. The only way this will ever be achieved is if one or more members of the core planning group advance a vision of what is ahead—possible worlds that the company will actually live in at some future date. This is not easy and the vision is often wrong, but creative people will stick their necks out and take a chance on being wrong. These individuals want to stand out and they are often known for a gush of ideas—many of which sound silly and eventually might be silly. But someone has to listen to these visionaries in a company. *It is the job of the core group to take this visionary outpouring and convert it into disciplined anticipation of the future.*

How does a company spot these visionaries if it is not obvious to the CEO? Think of the most creative *event* or critical incident in the past three years in the company. Not necessarily a patent application or a new product idea from a salesperson. Rather, think about a situation where someone truly came up with an innovative solution to an organizational problem. For example, an important or key insight at a staff meeting, or the modeling of a new behavior that the firm really needs like a quality promoting activity. Who was at the heart of this event? Get these people involved in the planning activities of the firm for design. Don't forget that it takes a creative person to find a creative person.

Design office of the future

Although group work in the form of teams, committees, or task forces is becoming ever more popular in the work place, group work is difficult. The concept of computer-supported cooperative work (CSCW) is a new field that has grown out of the collaborative technology movement. The CSCW concept is intended to accelerate and ease the process by which people reach consensus by using computers in innovative ways to assist decision-making in groups. Five years ago there was virtually no activity in this area, now there are millions of dollars being spent on promoting this idea.

There are at least five well-known laboratories that have pioneered in new ways to have a computer-assisted meeting (Fig. 12.1). These laboratories include Xerox's two Colab facilities, the MCC project Nick (now dismantled), EDS Capture Lab, at ICL The Pod, and the University of Arizona's Business Laboratory.

Computer-supported cooperative work hopes to capture the best of face-to-face meetings when groups have serious work to accomplish, and still take advantage of computer data bases and problem-solving tools that otherwise would only be available to individuals. In research funded by Arthur Anderson and others, the University of Michigan is experimenting with reconfigurable meeting rooms (Fig. 12.2). Each individual module can be networked to the other stations (Fig. 12.3), but the furniture of this technology allows the screen to disappear in the counsel so that it doesn't interfere with face-to-face communication. There appears to be no real limit to the configurations possible using this approach.

The Collaboration Technology Suite at the University of Michigan (Fig. 12.4) is focusing on the development of tools to facilitate same-time, same-place groups by "augmenting group work with information technology" (in Olson, 1989). This research has opened a whole new vista of possibilities for managing design that otherwise would not be.

- The "drawing board" where everyone gathered is back in a new form.
- The data base and tools needed to do first-class design are available but do not interfere with face-to-face interaction.
- The problem of incapability of computer-supported systems is gone with this approach.
- The lab can be located on neutral turf and not under the design or manufacturing roof.

The University of Michigan group has identified at least six "streams of information" that occur in face-to-face group meetings that often cause information overload for participants:

- *Progress* information
- *Agenda* tracking
- Monitoring and controlling *"turns"*
- Creating and editing the *work object* (that is, the design)

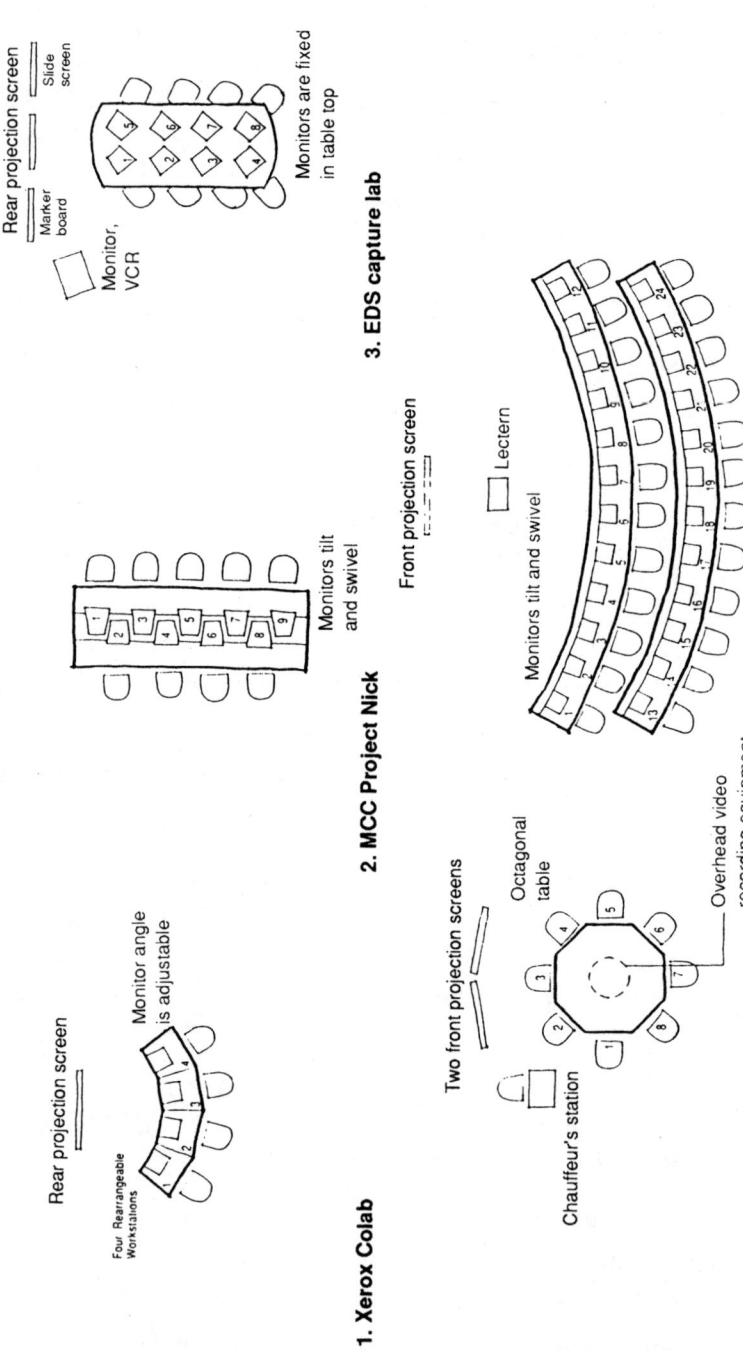

Figure 12.1 Existing CSCW facilities. (*Courtesy of Paul Cornell and the Human Factors Society.*)

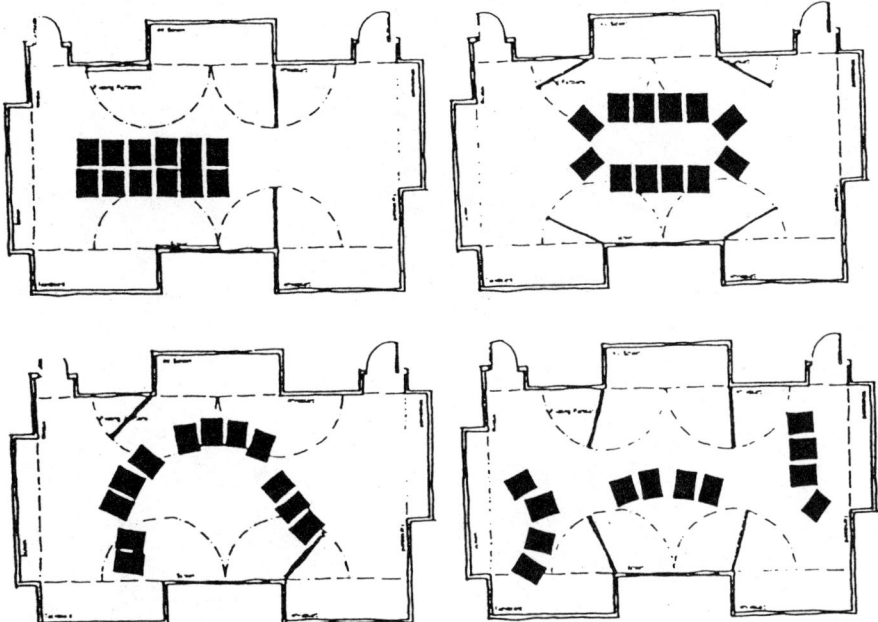

Figure 12.2 Divisible space. (*Courtesy of Paul Cornell and The Human Factors Society.*)

- Engaging in *activities* (for example, brainstorming)
- *Records*

A great deal of research effort is now focused on "*why* people collaborate, *what difficulties* they encounter, and *what aspect* of information technology may be able to help" (Olson, 1989).

We think some neutral turf location for a collaborative technology suite for design is what is needed in most design and manufacturing organizations. Applying the concept of computer-supported cooperative work will require close cooperation with universities and consultants as well as a disciplined anticipation for the future. Is your firm ready?

A new kind of creativity

Design is the essence of engineering. It is also the heart and future of many companies because the ability to compete effectively depends directly on the ability to design and produce what the customer wants. Creativity, or the act of creating, is the heart of design. Hence, in the long term, creativity is the lifeblood of successful companies.

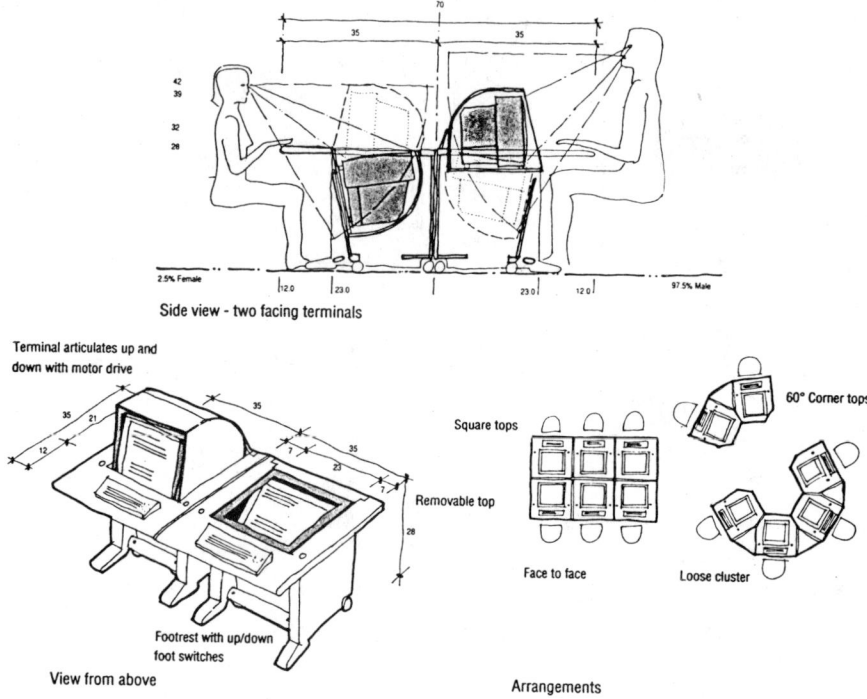

Figure 12.3 Articulating monitor workstation (*Paul Cornell and The Human Factors Society.*)

Historically, design creativity has been channeled along classical functional lines. In an R&D environment, for example, creativity is needed to invent and develop fresh ideas and product concepts that create new needs or satisfy existing needs in new or better ways. In product engineering and design, creativity is aimed at reducing cost or improving performance or both. Continuous improvement of existing designs by making relatively small changes to fix performance deficiencies, ease production difficulties, and improve manufactured quality also demands creative solutions. Innovative development and application of advanced manufacturing technology is yet another channel for creativity.

All of these channels for design creativity represent "traditional" creativity because of their tie to classical engineering functions. Although many opportunities for traditional creativity still exist and always will, in many industries, the "cream" has been skimmed and even minor advances require large amounts of effort (Fig. 12.5). What

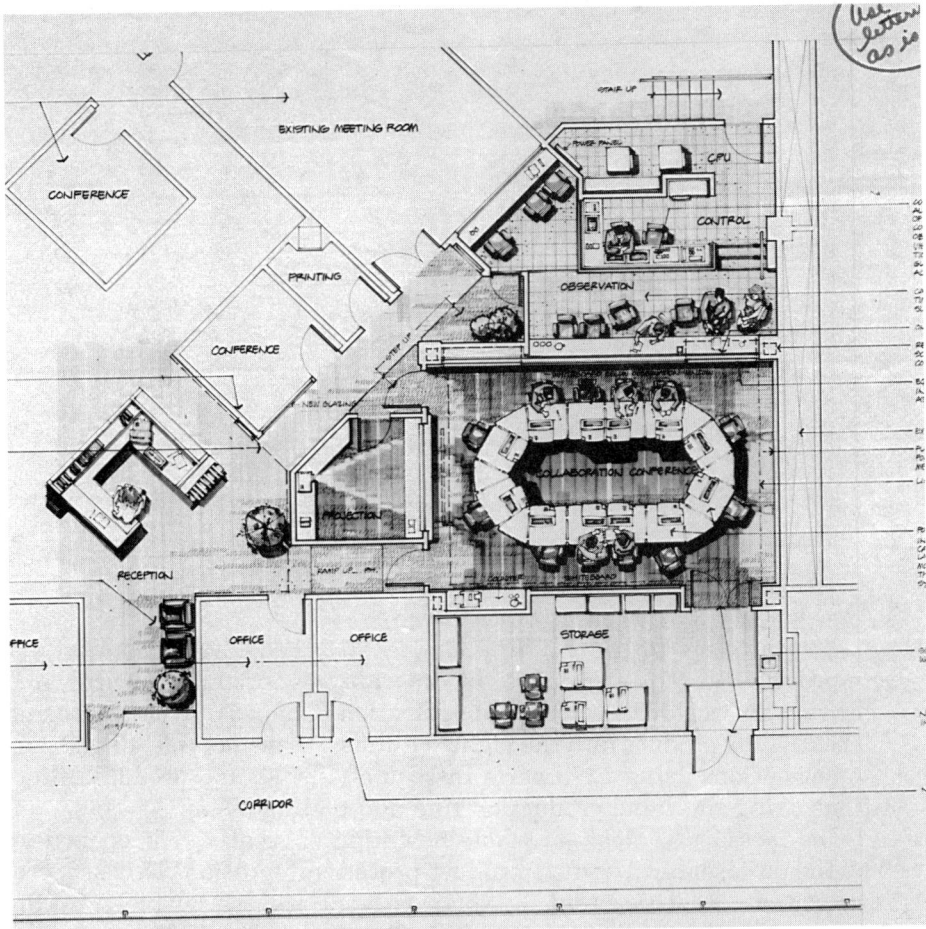

Figure 12.4 The Collaborative Technology Suite at the University of Michigan.

we have seen in most of the examples and case studies presented in this book is a "new kind" of creativity, born out of groups of people working *creatively* across functional lines to jointly solve several problems simultaneously.

By eliminating separate fasteners and consolidating parts, designers simplify product manufacture *and* improve load paths enabling higher performance and lower weight (Chap. 4). By focusing on an across the board change to Double Insulation technology, design teams redesigned an entire line of products to develop a "family" look, simplify the product offering, reduce manufacturing costs, automate manufacturing, standardize components, incorporate new materials,

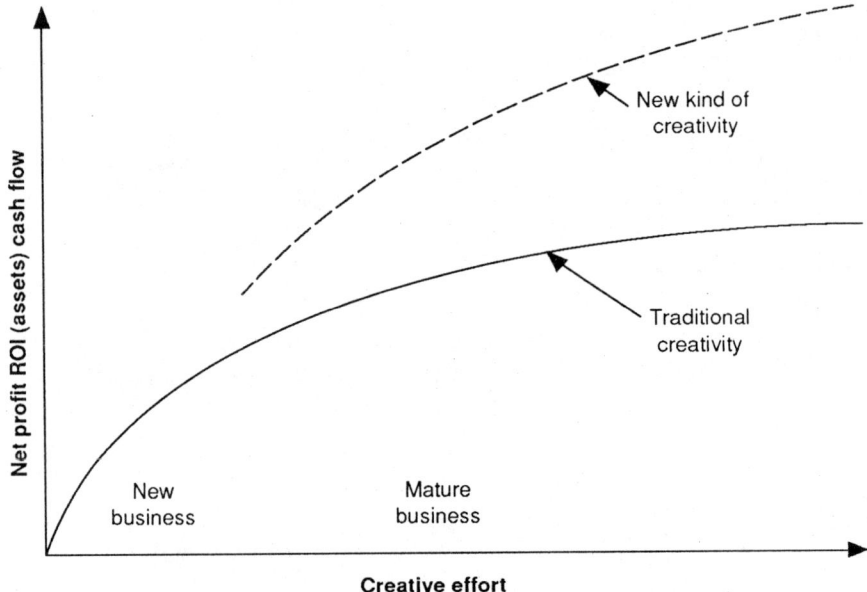

Figure 12.5 Impact of the new kind of creativity.

improve product performance, incorporate new product features, and provide for worldwide product specification (Chap. 5). By decomposing a particular product into the right "chunks," firms provide 40,000 customer options using 1275 parts instead of 28,000, and produce all options using the same production line and tooling (Chap. 7).

The message is clear—a whole new kind of creativity is opened up by the philosophical shift to product-process integration. This new creativity can greatly benefit the competitiveness of the business, taking over where the traditional creativity leaves off (Fig. 12.5), and at the same time, providing new avenues of opportunity for the more traditional channels of creativity. Essentially, the new kind of creativity involves seeing the total design—product, process, and people—in an entirely new light. It simultaneously and synergistically focuses creative attention on many facets of the business to produce improvements that are orders of magnitude better than those produced through separate creative efforts.

How can this new kind of creativity be stimulated and brought into play? We believe it will occur naturally as the philosophical shift takes place in the core company thinking. Once this shift has occurred, the following management actions, suggested by Jim Hughes, author of Chap. 7, can further catalyze the new kind of creativity within the company:

1. Focus on a product range in which your tactics, products, and processes can dominate, and organize internally to supply any variety within that range.
2. Migrate toward a customer-centered organization, focused on change-introduction both at the individual order level and at the new product level.
3. Continuously reduce time cycles for both current delivery and new product introduction.
4. Constantly up-grade information management systems to conduct routine operations, freeing personnel to handle nonroutine matters. Build useful cost management into the system.
5. Develop production facilities and systems for continuously matching the flow of materials and components. Include suppliers as partners.

Chapter 11 is a prime example of how to implement these ideas for stimulating this new kind of creativity. In Chap. 11, Bob Jones presented the view of this process of creativity enhancement from the perspective of an organizational development consultant at Cadillac Car Division of GM. Some believe so strongly in this perspective on the new design philosophy that they think the entire problem reduces to an organizational development or intervention problem. While we believe it is important, it is not the only factor that will determine success. But we do agree that creativity on cross-functional teams can be nurtured by expert intervention.

Design and the firm

At a recent conference on the strategic value of information in manufacturing, there were two days of presentations from practitioners and consultants. Three of these presentations were memorable from the standpoint of their depth, their lack of presumption, and their clarity. The companies involved are not important. The fact that none of the representatives of the these three companies used the term "CIM" is significant. They have all gone beyond CIM to a new plane of integrated manufacturing.

We believe one of the reasons that the "alphabet soup" of terms is finally diminishing is that general managers will not tolerate acronyms. Once general managers are truly involved in the change process—both reading grass roots and providing a vision—the link between technology, as represented in the core knowledge of design and manufacturing, and the business itself will be made. Only general managers experienced in the industry and the firm can make this link.

Although we detail the economic planning for design in the next section, it is worth noting that the *integration of the business plan with design and manufacturing is the essence of a successful corporate design strategy*. We need to reiterate the emergent, critical new role of the chief technical officer in this entire process. The resolution of conflicts between the management information function and manufacturing, in particular, will be very much dependent on the behavior of the chief technical officer as an arbitrator.

The tasks of top managers would not be complete unless they spent time on selecting middle managers—even in a dual ladder company. These people should have optimal stays in positions, not yanked too soon or delayed too long. The human resource officer of the firm needs to be involved in selecting members of the core team and implementation of strategies will depend on engineering mobility.

Finally, the concept of flexibility is now understood as a strategic issue in most leading-edge manufacturing firms. We are not restricting ourselves to just discrete parts manufacturers in industries like aerospace, automotive, appliance, and equipment. Five years ago, flexibility was not an issue in the chemical industry. Now with product diversification, it has become an issue for all strategic manufacturing managers. Further, flexibility can be used to posture and control suppliers, not just as a competitive weapon for meeting customer needs. If the supplier that makes the most profit on your business knows that you can migrate across the make-buy decision at will, your supplier costs will go down by merely getting into a product group with a flexible factory.

To summarize, Fig. 7.3, which is based on the experience at GE and any other firms, captures much of the general wisdom for general managers in this area. Simplified and focused main-line work flows, combined with well defined and closely coordinated interfaces between all basic activities of the business, result in an efficient order to delivery cycle, a smooth transition of new products from development into production, and inherent flexibility to be both proactive and responsive to change. Beyond that, it is essential to recall that a general manager's job does not stop with a vision of the future—it stops when strategies are implemented.

Economic Planning for Design

A major piece of unfinished business and a significant part of the puzzle for managers is the *economic planning of design*. We know of no outstanding, valid costing system available today that could be called a model for adoption in most manufacturing industries to guide the design process. The vast majority of systems available today do not in-form upstream decisions in design and do not adequately map the impact of these decisions downstream in the process. Not only are these costs difficult to estimate, the market implications of design decisions are subject to great speculation.

There are some early signs that this situation is changing. We present two examples of this trend. The first example is a cost system focused specifically on design issues. The second is representative of work to improve the general planning process for costs and benefits that involve new technology.

James Anderson, who is the director of advanced systems at Xerox Computer Services contributed a paper in the *Proceedings* of the National Association of Accountants (1988) entitled: "Unit Manufacturing Cost Tracking Systems at Xerox." The paper reports the shift at Xerox from cost accounting to cost management. This shift was not caused by new technology. Rather, Anderson says that Far East competitors "came in with lower prices, better deliveries, *and* (our emphasis) lower costs (implying better quality)."

Up until that time, Xerox had prided itself on how good their service organization was. The point driven home by this competition was that Xerox had never asked the fundamental question of why all this service was needed in the first place. If they sold a quality product, the product would not have to be serviced. At the same time this was happening, engineering and technology were fast becoming the largest cost categories for any product they sold, while direct labor was shrinking.

One of the points repeatedly illustrated in this book is that costs are not *determined* downstream in the product life-cycle. It is true that they are *incurred* downstream—and this is where most accounting systems are applied. This fact was embraced by Xerox and estimates were made that indicated that perhaps 30 percent of the cost of a product could be removed by excellent design and an additional 20 percent could be taken out by upgrading and changing manufacturing practices. Further, it seems clear to us that once a firm goes through this process, the way to capture benefit for being different is to learn from the process of doing it. *Once Xerox spreads this learning across its entire organization the benefits of the exercise are multiplied many-fold.* This becomes the next challenge.

Anderson says that as we change our design practices (for example, to eliminate parts, standardize components, adopt new materials, and simplify assembly), we also need to change our accounting systems. He contends that every new part we create costs $12,000 in administrative expenses (indirect cost of doing business). We should be reducing the number of parts and reexamining existing parts as the new engineering culture for design is adopted in a firm. *Target costs are key*. Target costs are not achieved by squeezing production. They are achieved through design and by standardizing the process of design so that parts can be sourced anywhere in the world.

Anderson cites three barriers that were encountered at Xerox that stood in the way of adopting a unit manufacturing cost (UMC) monitoring system:

- No common language
- Information scattered around the firm
- Complex international interactions to manage

The answer to solving these problems was to work hard to create a new system. At Xerox, unit manufacturing cost is the combination of landed direct material charges, and special charges, as well as material overhead and an allocation for tooling. The calculation of unit manufacturing cost (UMC) is given in Table 12.1.

Being able to calculate these costs involves accessing many distributed, heterogeneous data bases. The solution to this problem is not obvious. Accurate, timely data is not easily obtained under distributed

TABLE 12.1 Calculation Methodology

Business considerations	Material	Tooling	Labor
1. Landed Direct Mat'l. Cost Dom. freight and special charge rate	DMC DF&SCR		
2. Dom. freight and special charges Overhead rate	DF&SC = DMC × DF&SCR O/HR		
3. Overhead cost	O/HC = DMC × OHR		
4. Initial production tool cost		IPTC = IPTC ÷ #LB*	
5. Post line-balance tool cost		PLBTC = PLBTC ÷ #PLB**	
	1. DMC 2. DF&SC 3. O/HC 4. + IPTC 5. PLBTC		
Total:	UMC for the piece part		

*#LB = Number of Machines at Line Balance
**#PLB = Number of Machines after Line Balance
Note: To summarize, the sum of Direct Material Cost, Domestic Freight and Special Charges, Material Overhead Cost, Amortized Initial Production, and Post Line-Balance Tool Costs is the UMC for the piece part.

circumstances of this type. To migrate from one type of distributed system to a centralized data base is no small task and would cost several hundred million dollars at Xerox.

At the early stages of implementation, Anderson says that Xerox chose to

> build a product which would allow us to extract the information from all the mainline and end-user systems into one place, and then we would lift the data from that (Anderson, 1989).

A product delivery team was chosen to implement this solution and a UMC manager was selected to head the team.

For the UMC manager, a preoccupation with identifying cost reduction opportunities is paramount. If engineers keep configuration data up-to-date, this is an easier task. Above all, the target for costs is always the goal. In addition, "Forward Product Procurement buyers" (Anderson, 1989) at Xerox are the final link in the chain along with vendors or suppliers. These buyers bring new products to the firm and get quotes.

Both snapshot (standard cost if frozen today) and variance reports help focus the product delivery team effort. The impact at Xerox of installing the UMC tracking system has been primarily to consolidate information in a way that makes it much easier to access and use data. For example, off-line systems were eliminated. This saves time and avoids surprises downstream. Variances identify problems early in the design cycle and the integrity of the design is maintained across functions represented by the design team. Managers can realize target costs before products are launched at Xerox.

A second and more general approach to this problem of economic planning for design is represented by the work of Mike Burstein and his colleagues. For example, Burstein and Graham (1989) have dealt in great detail with the issue of strategic justification for integrated technologies like CAD/CAM, and outline a general approach which is reprinted in Fig. 12.6. In this approach, the general sequence of planning steps is outlined, but the idea is to rapidly go beyond this to detailed "market-based planning." Like the Xerox approach, it is focused on information and information systems needed to provide the decision-maker with the means to effectively link marketplace and manufacturing. In our approach, design is the critical link to manufacturing from the marketplace, but many alternatives are possible depending upon the goal.

In the general approach suggested in Fig. 12.6 for CIM strategic justification, the single most important step for our purposes is the relationship between distinctive manufacturing capabilities, which is often the missing link in a manufacturing strategy, and "manufacturing philosophies" as Burstein and Graham (1989) refer to them.

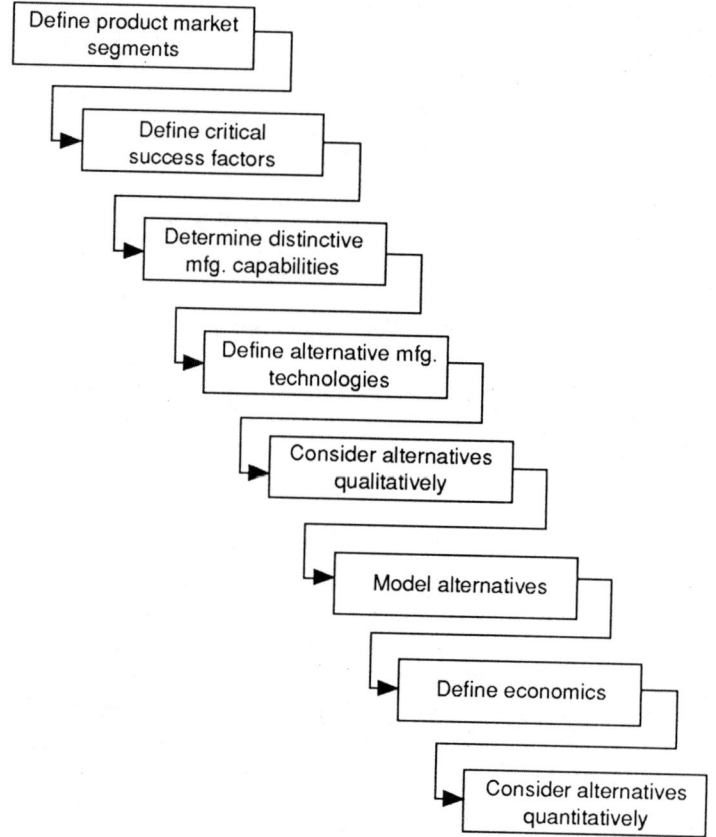

Figure 12.6 Simplified model of CIM strategic justification. (*Reprinted, by permission, from Manufacturing International 1990, vol. 1, p. 22, The American Society of Mechanical Engineers.*)

These manufacturing philosophies might include quality approaches (for example, Demming), material control methodologies like MRP and JIT, and of course, design philosophies like DFM, Group Technology, or Value Analysis. We would like to add human development philosophies to this list, while we are at it, because the latter is so essential to any change initiation and change sustaining effort.

Although Burstein and Graham distinguish between and qualitative and quantitative analysis steps in the process (fifth and eighth steps in Fig. 12.6, respectively), they also imply that the distinction between "tangible" and "intangible" is losing its meaning for manufacturing managers given the broad set of responsibilities these managers now assume. One of these new responsibilities is direct dealing with customers—both next in line and end customer.

As we have said earlier (for example, Chap. 3), one of the remaining challenges is the derivation of valid "upstream" predictors of design system outcomes. Measures of simplification and satisfaction of design team members are among these proposed measures.

One of the aspects of the design process that makes economic planning so difficult is that history shows that upstream costs are often not "real." A case illustration might serve to make this point. We know of an instance of a new component and technology that was being developed by a particular automotive manufacturer for use in passenger cars. To be first-to-market with this new feature on an existing model, the program was pushed ahead by one model year, in anticipation that competitors would be forced to follow in the next year's introduction cycle. In order to "crash" this program, the car company division had to reallocate engineering effort to the project, but did not rebudget this effort. The distribution of engineering time often does not match how engineers charge their effort in accounting systems. Therefore, *the attitude that costs are not real often prevails in design.*

Another tendency that dilutes the "reality" of costs is that there is often the perception on large, calculated risk projects, that money spent early in the project is less precious than money spent later in the project. As budget levels get lower, and "risk" or "buffer" pots of money are tapped in order to carry a project through, team members often become very nervous about (budget) performance.

While spending can often be controlled, time is often less controllable. This interaction between time, time into a project, and perception about progress toward the goal changes how monetary units are viewed. We know of several instances where initial planning efforts on a new product or a product and manufacturing system change project, have been expensed rather than carried or forced through an appropriate request sequence. Often, corporate budgets are used to fund these earlier project stages. This adds to the perception that not all money is equal. "Our money is more equal with fewer strings attached" often describes the feelings of participants in these significant new program efforts. At least one case we are intimately familiar with was justified by working backward from target ROI to estimate costs. Were all of these cost reevaluated? Hardly.

Understandably, corporate technology managers are quite interested in increasing the return on R&D investments. One attractive way of doing this involves the sharing of technology across the corporation and spreading the word on what was learned in the process of launching a new product-process system. GE is a company that appears to do a very reasonable job of capturing the essentials of each divisional learning experience and sharing these across the corporation. The methods they use include both formal programs and seminars, the GE internal report system, and a well developed and nurtured informal communications network. Internal consultants are no small part of the mechanism that maintains this corporate information network.

An important feature of any economic planning process is the ability to determine what a reasonable expectation of a progress function or learning curve might be on a project. After a new product-process system is released to manufacturing, there is a two-stage process of start-up and ramp-up initiated. We have observed on several occasions that the slope of the curve of start-up and slope of the curve of ramp-up, as well as the beginning and end points of these periods, is not like the prediction. The prediction and actual levels determine benefit streams in a marketplace that is seldom forgiving. It is not surprising that several groups of serious researchers are revisiting the topic of learning curves in their projects (for example, Linda Argote at Carnegie-Mellon and John Bessant at Brighton Polytechnic in the United Kingdom). The more we know about this learning curve phenomena under conditions of uncertainty, the better our economic planning will be.

Dick Place of the Ford Motor Company presented a talk entitled "Lessons from Japan: Learned and Not Learned," at a recent Automotive Seminar Luncheon at the University of Michigan (October 9, 1989). Ford's partner in Japan, Mazda, asked their suppliers to reduce costs 25 percent in 30 months during the 1980s. The target of 5 percent reduction every six months was actually *exceeded* by these Mazda suppliers. Could your suppliers meet this progress function goal? What have you done to help them meet targets like this to reduce costs to a world standard low?

According to Dick Place, Ford is trying hard to become a "customer driven company" using quality function deployment (QFD) and similar philosophies. What does it mean to be customer driven? It clearly is *not* the old slogan that the customer is always right. Rather, to us it means translating customer needs into internal requirements. In a way, the ultimate success is measured in how well you satisfy yourself that customer needs are met better by you than by your competitors. If your competitor knows your customer better and knows himself better, he will beat you. Knowing what your competitor is going to do actually comes in a close third in this approach and philosophy. The key is, know your customer and yourself first, then your competitor.

Where partnerships are concerned, Dick Place also has a message based on Ford's experience with Mazda. In order to be successful, at least two important criteria need to be satisfied:

1. The partnership yields benefits to *both* partners *equally*.

2. The partners respect each other as equals.

The second criteria has an important implication for economic planning for design. Ford and Mazda agree to do things *before* they are done. Only mutual respect can make that happen.

At the recent Fifth Annual National Quality Forum, the American Society for Quality Control released the results of a Gallop survey (*Wall Street Journal,* 1989). The findings indicated that "U.S. companies now overwhelmingly feel their fellow Americans—and not the Japanese—present the greatest competition on quality." Senior executives attending the Forum *objected* strenuously and broadly to these survey results. David Luther, Senior Vice President and Corporate Director of Quality at Corning, Inc., reportedly said: "They're totally mad. How can they say that Americans are our stiffer competitors," when the Japanese "are absolutely killing us in a number of areas." GM president, Robert Stempel, said that the survey probably reflects American progress on quality but that "we haven't taken our eyes off Japan."

The point of reviewing this discourse on quality is to highlight the importance and role that perceptions of quality, and therefore, cost play in economic planning. Not only are we still concerned about quality, we must realize that the rules of the game have changed. We have to be able to move *more* than one chess piece on the board at the same time to meet and better competition. Which pieces we move and our strategy for these multiple moves will determine the outcome of the game. We argue in this book that an essential part of this new multiple move strategy is a shift in design philosophy.

If you are preoccupied with estimating existing costs, you've essentially missed a major point of this book. Like the new creativity, we believe that once this philosophical shift has occurred, the organization will gradually begin to focus in on true cost. These true costs form a much better basis for understanding the true resources and capability of the firm.

Summary

This book is an attempt to show that there is a bridge to the future in manufacturing. The particular bridge we construct is one built on several principles beginning with a philosophical shift in a firm from one approach to design to another. It makes little sense to talk about the details of this bridge to the future until one has at least a gut feeling for what this philosophical shift represents. Stated in one sentence, this *new philosophy is that designs that are enduring are simply pro-*

duced by groups of people working together providing their best contribution to a whole. This whole is remarkably simpler than what came before and easier to implement. This design requires fewer changes in the future and meets target costs before it even goes into production. When the design has to be changed, it is changed to provide true incremental benefit to the strategic unit, not to make up for mistakes or to capture some small, marginal return. Because the design is inspired by general manager vision, it is easy for everyone to grasp company-wide. General managers do not call for strategies or designs that are not easily understood globally. In order to capture the multiplier benefit of being innovative, what is learned each time designs are envisioned and executed, is shared corporately. No one can imitate or reverse-engineer this residue of the learning experience. This is the essence of what is unique about a company culture that has competitive advantage.

Cases in this book illustrate various aspects of implementing this philosophical shift in its various manifestations. Each company has its own version of this new truth. But there are similarities in these approaches that we would like to emphasize.

The philosophical shift comes first. This is difficult and takes time and cannot be expected to happen overnight. Some enduring core of the organization has to embrace the need to change and the urgency to move to a new standard of practice before the shift will occur. All of our cases share this common denominator.

Second, as change occurs, people involved will acquire a new language—common to a changing core that is representative of the attitude shift occurring in the process. The attitude that is at the heart of the change is the value placed on the design approach that anticipates the future. Joint anticipation is disciplined anticipation—by definition.

Third, our cases show that learning can be accomplished and it is stimulated and supported, even maintained, by everyone taking responsibility for human resource development in the company. This includes taking charge of their own destiny as well as mentoring the development of others. Not everyone will be involved equally, but everyone will be involved.

Fourth, and finally, a new philosophy for design can have many starting points in many different settings, but all companies, regardless of size and industry, seem to get around to convincing themselves that there are a few essentials to this new approach. These essentials include respect for the other person's voice in the consensus-reaching process, regardless of status or past performance, and understanding that partners outside the firm have a contribution to make to learning, and that shared destiny is a value that all must embrace before real progress toward the future can be made. The shared learning experience becomes part of the complete ring of the wood grain of growth, texture of strength in competition, and the means by which we ultimately satisfy ourselves as we satisfy our customers. Our designs endure as our culture sustains.

References

Anderson, James R., "Unit Manufacturing Cost Tracking Systems at Xerox," *Proceedings*, National Association of Accountants, Montvale, NJ, 1989, pp. 185–192 (and presentation foils).

Bennet, Armanda, "The Survey Sounds too Bad to Be True," *Wall Street Journal*, Tuesday, October 10, 1989, p. B1 (Reviews the Quality Forum of the same year).

Burstein, Michael C., and Graham, Pearson, "Strategic Justification of CIM: A Systemic, Market-Based Approach for Plant Determination of Distinctive Manufacturing Capabilities," working paper submitted for the MI'90 Conference Proceedings of the ASME, 1989.

Olson, Gary M., "The Nature of Group Work," in *CSCW: Evolution and Status of Computer Supported Work*, Paul Cornell, Robert Luchetti, Lisbeth A. Mack, and Gary M. Olson, Technical Report no. 25, Cognitive Science and Machine Intelligence Laboratory, Univ. of Michigan, Ann Arbor, August 1989.

Place, Richard, "Lessons from Japan: Learned and Not Learned", Automotive Seminar Luncheon, Office for the Study of Automotive Transportation, University of Michigan, Ann Arbor, October 9, 1989.

Index

A. B. Chance Company, 201, 260
A. T. Kearney, 40
Activity/transaction based accounting, 9, 111
Activity value, 172–173
Advisory council, 180–181
Anderson, James, 269–271
Anticipating requirements, 55
Applied Computer Solutions, Inc., 216
Argote, Linda, 274
Axiomatic design, 94–95

Barnard, Chester, 42
Barney, Jay, 26
Beatty, Carol, 13
Bendell, A., 103
Bessant, John, 274
Blache, Klaus, 133, 146n, 147, 148n
Black & Decker, Inc., 22, 92, 98, 117–131
  double insulation, 119–122, 125–130, 265
Boeing Company, The, 16
Boothroyd, Geoffery, 108, 215
Boothroyd-Dewhurst DFA method, 108, 215
Boston University Manufacturing Roundtable, 22n, 50
Bradyhouse, Richard, 92, 99
Brunswick Corporation, 79n
Building block parts, 96
Bureau of Labor Statistics, 51
Burget, Phillip, 53
Burnout, 49
Burns, T., 44
Burstein, Michael, 271–272
Bussey, John, 53

CAD (see Computer-aided design)
CADAM, 195

CAD/CAM, 13, 42, 62, 173, 259, 271
CAD/CAM/CAE, 89, 91, 92, 207
CAE (see Computer-aided engineering)
CAM (see Computer-aided manufacturing)
Champion, 245–246
Change:
  control boards, 212
  design for, 103–106
  engineering, 11, 24, 33, 235
  introduction organization, 168
  in philosophy, 227
  organizing for, 160–170
  oriented organization, 162
  process, 245
  survival, 118–125, 226–227, 258
  two classes of, 161–162
Chimneys of power, 28
CIM (see Computer-integrated manufacturing)
Clark, Kim, 42, 43, 54
Clausing, Don, 98n
Communication, 180
Competitor analysis, 130–131
  (See also Reverse engineering)
Computer-aided design (CAD), 21, 47, 56, 92, 106, 107, 189, 203–204, 212
Computer-aided DFM, 111
Computer-aided engineering (CAE), 207, 209–210
Computer-aided manufacturing (CAM), 204–205
Computer-integrated manufacturing (CIM), 25, 42, 189, 205–207, 267, 271
Computer numerical control (CNC), 211
Computer-supported cooperative work (CSCW), 260–261
Component design engineering (CDE), 47

279

Conceptual design, 83–85, 91–92, 209, 234
Concurrent engineering, 22, 86–87
  (*See also* Design integration; Overlapping approach; Parallel engineering; Simultaneous engineering)
Continuous improvement:
  design-right-first time, 241
  gate procedure, 238–240
  key concept, 141
  link between design and manufacturing, 138
  product development and improvement teams (PDITs), 249
  simultaneous engineering teams, 251
  system for, 146–147
Controllable factors, 101
Cookbook solutions, 52
Core knowledge, 176
Core team, 259
Core technology, 171
Corning, Inc., 3n, 275
Cost reduction, 125–127
Cost systems, 9, 111, 269–275
Couglan, Paul D., 223
Cox, Jeff, 51
Crawford, Merle, 3n, 4, 4n
Creativity, new kind, 263–267
Crisis response, 118–125, 226–227, 243
Cross-functional integration, 236–237
Cross-training, 44
Culture:
  anthropological view, 44
  blending of, 44
  bridging the gap, 55
  company size, 38
  complexity, 33
  culture gap, 46
  demographics, 39, 51
  design-manufacturing interface, 25
  engineering education, 30
  of engineers, 21
  imitation of, 5
  manufacturing engineers, 36–42
  manufacturing science, 41
  organizational, 25–27
  orientations, 28–31
  of product engineers, 31–36
  workplace, 29, 37
Customer driven, 274
Customer focus, 174–175

Data General Corporation, 34

Davis, Stanley M., 27
Dean, James W., 26
Deere and Company, 22
Dertonzos, et al., 30
Design:
  deadlines, 31
  economic planning of, 268–275
  freedom, 84
  fundamental question of, 80
  iterative nature, 80–82
  one-pass, 189–191
  phases of, 81
  (*See also* Product development stages)
  process driven, 100
  variation tolerant, 102
Design and the firm, 267–268
Design cycle time:
  A. B. Chance, 205
  administrative charges, 68
  automotive engines, 149–150
  benefits of, 170–173
  Cadillac Eldorado, 244
  definition, 170
  evaluation of contributed value, 172–173
  gate procedure, 229
  IBM model 5363, 195, 198
  metal bending, 3
  1955 Chevrolet, 8
  P-51 Mustang, 7
  Quad-4 engine, 9, 134
  reflections on, 258
  targets, 193
  Xerox Corporation, 3
Design engineer, 31, 47, 50, 51, 58–59
Design for analysis, 107–108
Design for assembly:
  benefits at A. B. Chance Company, 215
  Boothroyd-Dewhurst method, 108, 215
  design process, 93, 210
  GE Salisbury project, 181
  toolkit, 111
  training, 108, 112
  use of, 91
Design for automation, 227, 229
Design for change, 103–106
Design for flexible manufacture, 106–107
Design for manufacture:
  approaches, 93–112
  basic precepts, 81
  B24 Liberator, 5
  concepts, underlying, 80–89
  definition, 79

Design for manufacture (*Cont.*):
  design guidelines, 206
  disciplined management approach, 230
  Harmony telephone (Table 10.1), 228
  narrowing of design choices, 80–81
  philosophy:
    approaches, 93
    definition, 85
    design for assembly, 108
    design tools, 215
    flexible manufacture, 106
    need for, 206
    new opportunities, 107
    product descriptions, 85
    reviews, 210
    standardization, 122
    system design, 102
    toolkit, 110–112
    tools, 215
    training, 56, 68, 74, 112
Design for quality, 101–103
Design integration, 85–88
  (*See also* Design-manufacture integration)
Design iteration, 82, 213
Design life, 56
Design management principles, 257–268
Design-manufacture integration:
  actions, five key, 56–57, 70–73
  audit tools, 64, 68, 70–74
  cautions, 69–70
  challenges, 49–50
  closing the culture gap, 46
  conceptual design, 234
  design process, 201–221
  early manufacturing involvement, 192
  evaluation, 67
  in-process measures, 67
  location of manufacturing engineers, 120
  management perspective, 42–49
  measure of satisfaction, 64–69, 73–74
  novel organization structures, 61–63
  optimum level, 257
  organic management structures, 44
  outcomes, 57–58
  policies, 65
  product development approach, 85–88
  results, 122
  satisfaction with, 64–66, 69
  scale development, 73–74
  specialization, 45
  structuring for, 59–64

Design-manufacture integration (*Cont.*):
  success, measurement of, 66–68
  suppliers and customers, 199
Design-manufacture interface, 25, 26, 32, 33–34, 42–49
Design-manufacture paradox, 55–56
Design objectives, 93, 97–98
Design office of the future, 260–261
Design philosophy, 213–220, 230, 272
Design procedures, 92
Design process:
  at A. B. Chance Company, 208–212
  CIM, 205–207
  disciplined approaches, 92–93
  hard to control factors, 103–104
  improvement, 89–93, 260
  integration, 201–221
  iterative model of (Fig. 4.3), 82
  management of, 22–25, 207–208
  successive approximation analogy, 90–91
  top management role, 267–268
Design responsibility, 198
Design review:
  check points, 207
  documentation, 209
  engineering, 211–212
  event-driven, 237
  gates, 231–232
  IBM model 5363 design, 197
  manufacturing sign-off, 57
  at Northern Telecom Canada, 229
  producibility engineering, 37
  project performance, 236
Design simplification:
  GE Salisbury project, 181–182
  guidelines for, 95–98
  IBM model 9404 design, 191
  new creativity, 265–266
Design specification, 82, 177, 207, 227, 234–235
Design team, 208, 211
  (*See also* Team approach; Teams)
Development cycle, 53, 234–236
Development lead time, 19
Dewhurst, Peter, 108, 215
DFA (*See* Design for assembly)
DFM (*See* Design for manufacture)
DFx tools, 108–109
Direct numerical control (DNC), 204
Disciplined anticipation, 4, 260, 276
Disciplined design process, 89–93, 260
Disciplined management approach, 223, 230

282 Index

Dixon, John, 82
Dixon, Robb J., 51
Double insulation (*see* Black & Decker Corporation)
Drafting time, 213
Dyer, Davis, 45
Dynamic simulation analysis, 203, 204, 213, 214, 219

Early manufacturing involvement (EMI), 22, 187–199
Eastman Kodak Company, 3n
Ebert, Ronald J., 18
Ebner, Merrill L., 21n
Economies of scope, 21
Electric field analysis, 210
Electronic data interchange (EDI), 189
Electronic Data Systems, 261
EMI (*see* Early manufacturing involvement)
Engineering change, 11, 24, 33, 235
Engineering charge-backs, 69
Engineering education, 30
Ettlie, John E., 21n, 53, 54, 56, 61, 133

Failure mode and effect analysis (FMEA), 111, 135, 210
Fault tree analysis, 210
Finite element analysis (FEA), 204, 210, 213
Fixturing, flexible, 107
Fixturing, jigless, 14
Flexible manufacture, 100, 106–107
Flexible manufacturing systems, 21, 54, 62
FMC Corporation, 53
Ford Motor Company, 5, 22, 274–275
Fujimoto, Takahiro, 43
Functional integration, 233–234, 236–238
Functional requirements, independence of, 94
Funnel process, 83

Gate procedure, 223–242, 260
General Electric Company (GE):
    Salisbury project, 181–184
    technology transfer, 273–274
General Motors Corporation:
    Cadillac Automobile Company, 243, 258, 267
    Chevrolet-Pontiac-Canada Group, 53
    Delta engine plant, 133, 144, 147–148

General Motors Corporation (*Cont.*):
    1955 Chevrolet, 8–12
    Quad-4 engine, 67, 133–157
    quality, 275
Ginn, Martin E., 29, 30, 43
Goldratt, E. M., 51
Gordon, John R. M., 13
Graham, Pearson, 269–272
Group technology, 111
Guidelines, design simplification, 95–98
Guidelines for purchased parts, 137

Hancock, Walter M., 32, 33
Handy Company, 18
Harmony telephone project, 226–229
Harris, Roy J., 16n
Haubein, H. Dennis, 201
Hayes, Robert H., 42, 43
Hewlett-Packard, 22
Historical precedents, 5–12
Hitachi Seiki, 41, 42
Holusha, John, 3n
Huber, Robert F., 142n, 148
Human resource development, 49
Huthwaite, Bart, 95, 98

IBM Corporation, 22, 187–199, 258
Improvement (*see* Continuous improvement; Design process)
Information:
    availability, 81, 148
    content, minimization of, 94–95
    streams of, 261, 263
Innovation, managing of, 12–14

Jaikumar, R., 41
Japanese advantage, 54
Job rotation, 57, 68, 74
Johne, Axel, 12
Jones, Robert F., 243, 267
Judgment capture, 18

Kidder, Tracy, 34
Koenig, Daniel T., 36, 50
Koska, D. K., 40
Krampert, Jim, 17

Lawrence, Paul R., 45
Lawrence, Peter, 3n
Layout designer, 47
Leader, 175–176, 215
Learning curves, 274
Learning from others, 246
Lehnerd, Alvin P., 117

Liker, Jeff K., 32, 33
Lockheed Corporation, 3, 22
Lorsch, J. W., 45
Luther, David, 275

Machine vision, 99
Majchrzak, Ann, 66
Majerns, Clyde D., 18n
Management procedures, 242
Management systems, 44, 46
Management tool, 241
Manager characteristics, 24, 35
Manufacturing designers, 195–197
Manufacturing engineering philosophy, 214–215
Manufacturing engineers, 36, 50, 51, 58–59
Market share, 129
Marketing review, 209
Mazda, 274
Metcalf, George F., 8n
Miles, Donald L., 133
Minnesota Mining and Manufacturing Company (3M), 22
MIT Commission on Industrial Productivity, 30
Modular assembly, 192
Modular building block approach, 182
Modular design, 104
Monsen, William H., 187
Motivation, 46

Nadler, David, 19
Nanni, Alfred J., 51
National Academy of Engineering, 51
1955 Chevrolet, 8–12
Northern Telecom Limited, 223–242, 258, 260
Northrop Corporation, 22
Nowak, Steven, 133

Olson, Gary, 261, 263
Operations integrator, 43
OPT principle of manufacturing, 51
Order-to-delivery cycle, 169, 182
Organic management structures, 44
Organization design, 246–249
Organizational adaptation, 46
Organizational change, 14–20, 26, 251
Organizational culture, 25–27, 244
Organizational development, 244, 246
Organizational development consultant, 267

Organizational leadership, 244–245
Organizational structure, 54, 57, 59–63, 68, 91, 166–170, 259–260
Organizing for change, 160–170
Outcomes, 212–218
Overlapping approach, 43, 54, 86

Parameter design, 102
Parallel engineering, 53
Part count reduction:
 A. B. Chance Company, 215–216
 audit metric, 13–14, 109–110
 guidelines, 95–96
 GM Quad-4 engine, 135
 GE Salisbury project, 181
 NT Harmony telephone (Table 10.1), 228
 Sunbeam Appliance Company, 130–131
Patikonda, Mohan V., 22n
Performance measurement, 51
Personnel movement, 57, 68, 74
Philosophical shift, 257–260, 275–276
Place, Richard, 274
Plant organization, 144–145
Port, Otis, 3n
Predictive measures, 13
Principles, new, 257–268
Problem solving, 28, 45, 213
Process-driven design, 99–100
Process focus, 192–195
Producibility, 37, 109–110
Product descriptions, 85
Product design, 104–105, 117–118
Product development, impact on, 129–130
Product development advisory committee (PDAC), 202
Product development and improvement teams (PDITs), 248–249
Product development stages, 234–236
Product development teams (PDTs), 249
Product focus, 174–175, 191–192
Product-process change, 160
Product-process development, 185, 258
Product program, Quad-4, 135–138, 149–152
Product technology, 170–172
Professional obsolescence, 48
Professional orientations, 28
Professionalization, 28, 29
Project, frequent short cycles, 229–231

284 Index

Project approach, 173–181
Project leader, 175–176
Project management, 173–181, 223–242
Project management philosophy, 260
Prototyping, 198, 211, 213–214

Quad-4 engine, 133–157
Quality, 144, 146–147
Quality function deployment (QFD), 111, 177, 274

Raelin, Joseph A., 28
Readaptive process, 45, 49–50
Reifeis, S. A., 56
Reliability, 134, 137
Reorganizing, cautions for, 69
Reverse-engineering, 98, 104, 134, 177, 276
Review process, 210–212
Reviews, gate, 231–233
Robust design (see Design for change; Design for quality)
Romano, J. D., 40
Rose, M. F. (Bud), 79n
Rosenthal, Stephan R., 21
Rubinstein, Albert H., 30, 43
Rude, Dale E., 18n

Salisbury project, 181–184
Sasser, Earl, 3n
Sease, Douglas R., 53
Self-qualification, 135
Serial approach, 22, 85
Shapero, Albert, 49
Shinizu, M., 107
Simmons, Melvin K., 82
Simultaneous engineer, 139
Simultaneous engineering:
  continuous improvement, 251
  definition, 247
  design cycle, 258–259
  design integration, 86–87
  implementation of, 243
  implementation workshop, 249–250
  learning, 252
  new approach, 22, 55
  positive effects, 252
  Quad-4 engine project, 134
  role of top management, 247
  steering committee (SESC), 245–246, 248
  structure for, 246–249
Snelson, Patricia, 12

Society of Manufacturing Engineers (SME), 40
Socio-technical systems (STS), 46, 49
Specialization, 28, 32, 45
Stalker, G. M., 44
Standardization, 95, 99, 119, 122, 128–130
Standardization and rationalization (S&R), 98–99
Statistical problem solving, 111, 210, 214, 215, 218–219
Statistical process control (SPC), 111, 144, 218–219
Steering committee (see Simultaneous engineering)
Stempel, Robert, 275
Stoll, Henry W., 79, 238
Success factors, 176–177
Suh, Nam P., 94
Sunbeam Appliance Company, 120, 130–131
Supplier relationships, 142, 143, 169, 259–260
Supplier selection, 139–142
Supplier survey, 153–156
Suppliers, Quad-4, 139–144
Suppliers, reduction of, 135, 139
Suri, R., 107
Susman, Gerald I., 26

Taguchi, Genichi, 101, 102
Taguchi methods, 102, 103, 111
Taylor, James, 47
Team approach, 88–89
  communication, 180
  company-wide effort, 202
  design for assembly, 108
  design guideline, 206
  disciplined management, 230
  flexible manufacture, 106
  institutionalization, 252
  knowledge growth, 112
  leader, appointment of, 227
  leader, characteristics of, 175–176
  outsourcing, 134
  process-driven design, 100
  product development teams, 142, 143
  purchased parts, 137
  selection of team, 207
  team makeup, 177–180, 227
Team building, 147–148
Teams:
  choice of leadership, 69–70

Teams (*Cont.*):
  composition, 63–64, 68
  cross-functional, 68
  "daytime family," 24
  design/build, 18
  forming of, 34
  hourly employee involvement, 62, 70
  leader, 215
  observations, 148–149
  organization, 215
  project, 177–180
  satisfaction with, 56
  structure, 208–210
  task force, 24
  tiger, 53
  types, 61
Teamwork, 226–229
Technology transfer, 10–11, 134, 152, 273–274
Testing, 192, 198, 235
Thomson, M. F., 137n, 138n, 142, 144n
Time-based competition, 23, 161
Time to design (*see* Design cycle time)
Timken, 3n
Tjosvold, Dean, 19n
Tolerance design, 102
Tools, 108–109, 110–112
Training:
  computer-aided design, 203, 205, 212
  continuous improvement, 141
  design cycle reduction, 213
  design for assembly, 108
  design for manufacture, 56, 57
  statistical methods, 148, 218–219
  support activities, 169–170
Trist, Eric, 46

Tushman, Michael, 19

Uncertainty, 83–85, 241, 274
Unit manufacturing cost system, 270–271
University of Arizona, 261
University of Michigan, 261

Value engineering, 111, 120, 141, 227, 229, 235
Variation analysis, 101–102, 204, 210, 211, 214–216
VDI Society, 92
Vehicle system management teams (VSMT), 250
Verification in product and process, 144, 146–147
Vertical integration, 127–129
Vision, 244–245, 260, 276
Vollmann, Thomas E., 51

Wells Fargo, 4
Wheelwright, Steven, 3n, 43
Whinston, T. G., 46
Whitney, Daniel E., 13, 53
Wood, Albert R., 223
Work flow, organization based on, 91, 166–170
Work flow analysis, 162–166
Workshops, 250

Xerox Corporation, 3, 261, 269–271

Yasuhara, M., 94

Z-axis assembly, 96, 227
Zilog, Inc., 47

ABOUT THE AUTHORS

JOHN E. ETTLIE is the Director of the Office of Manufacturing Management Research, and Associate Professor of Operations Management at the School of Business Administration, University of Michigan. His previous positions include Senior Researcher for the Industrial Technology Institute in Ann Arbor, Michigan, and Assistant Professor of Management at the University of Illinois. Dr. Ettlie is widely published in magazines and professional journals.

HENRY W. STOLL is currently Technical Director for Design Technology at Square D Company's new Corporate Technology Center. His previous positions include Manager of Design for Manufacture at the Industrial Technology Institute in Ann Arbor, Michigan, and Professor of Mechanical Engineering at the University of Wisconsin—Platteville. A registered professional engineer and author and coauthor of many professional publications, Dr. Stoll is a member of the American Society of Mechanical Engineers, the Society of Manufacturing Engineers, and the American Society for Engineering Education.

TS 171.4 .E87 1990

Ettlie, John E.

Managing the design-
manufacturing process

DISCARDED

JUN 30 2025